OXFORD ENGINEERING SCIENCE SERIES

General Editors

J.M. Brady

C.E. Brennen

W.R. Eatock Taylor

M.Y. Hussaini

T.V. Jones

J. Van Bladel

RECENT TITLES IN THE OXFORD ENGINEERING SCIENCE SERIES

10 P. HAGEDORN: *Non-linear oscillations* (Second edition)
13 N. W. MURRAY: *Introduction to the theory of thin-walled structures*
14 R. I. TANNER: *Engineering rehology*
15 M. F. KANNINEN and C. H. POPELAR: *Advanced fracture mechanics*
19 R. N. BRACEWELL: *The Hartley transform*
22 C. SAMSON, M. LeBORGNE, and B. ESPIAU: *Robot control: the task function approach*
23 H. J. RAMM: *Fluid dynamics for the study of transcopic flow*
24 R. R. A. SYMS: *Practical volume holography*
25 W. D. McCOMB: *The physics of fluid turbulence*
26 Z. P. BAZANT and L. CEDOLIN: *Stability of structures: elastic, inelastic, fracture, and damage theories*
27 J. D. THORNTON: *Science and practice of liquid-liquid extraction* (Two volumes)
28 J. VAN BLADEL: *Singular electromagnetic fields and sources*
29 M. O. TOKHI and R. R. LEITCH: *Active noise control*
30 I. V. LINDELL: *Methods for electromagnetic field analysis*
31 J. A. C. KENTFIELD: *Nonsteady, one-dimensional, internal, compressible flows*
32 W. F. HOSFORD: *Mechanics of crystals and polycrystals*
33 G. S. H. LOCK: *The tubular thermosyphon: variations on a theme*
34 A. LINAN and F. A. WILLIAMS: *Fundamental aspects of combustion*
35 N. FACHE, D. DE ZUTTER, and F. OLYSAGER: *Electromagnetic and circuit modelling of multiconductor transmission lines*
36 A. N. BERIS and B. J. EDWARDS: *Thermodynamics of flowing systems: with internal microstructure*
37 K. KANATANI: *Geometric computation for machine vision*
38 J. G. COLLIER and J. R. THOME: *Convective boiling and condensation* (Third edition)
39 I. I. GLASS and J. P. SISLIAN: *Nonstationary flows and shock waves*
40 D. S. JONES: *Methods in electromagnetic wave propagation* (Second edition)
42 G. A. BIRD: *Molecular gas dynamics and the direct simulation of gas flows*
43 G. S. H. LOCK: *Latent heat transfer: an Introduction to fundamentals*
44 C. E. BRENNEN: *Cavitaiton and bubble dynamics*
45 T. C. T. TING: *Anisotrophic elasticity: theory and applications*
46 K. ISHIHARA: *Soil behaviour in earthquake geotechnics*
47 A. CHANDRA and S. MUKHERJEE: *Boundary element methods in manufacturing*
48 J. WESSON: *Tokamaks* (Second edition)
49 S. ARIMOTO: *Control theory of non-linear mechanical systems: a passivity-based and circuit-theoretic approach*
50 R. W. EASTON: *Geometric methods for discrete dynamical systems*

Geometric Methods for Discrete Dynamical Systems

Robert W. Easton

New York   Oxford
OXFORD UNIVERSITY PRESS
1998

Oxford University Press

Oxford New York
Athens Auckland Bangkok Bogota Bombay Buenos Aires
Calcutta Cape Town Dar es Salaam Delhi Florence Hong Kong
Istanbul Karachi Kuala Lumpur Madras Madrid Melbourne
Mexico City Nairobi Paris Singapore Taipei Tokyo Toronto Warsaw

and associated companies in
Berlin Ibadan

Copyright © 1998 by Robert W. Easton

Published by Oxford University Press, Inc.
198 Madison Avenue, New York, New York 10016

Oxford is a resistered trademark of Oxford University Press

All rights reserved. No part of this publication may be reproduced,
stored in a retrieval system, or transmitted, in any form or means,
electronic, mechanical, photocopying, recording, or otherwise,
without the prior permission of Oxford University Press.

Library of Congress Cataolging-in-Publication Data
Easton, Robert W., 1941–
Geometric methods for discrete dynamical systems / by Robert W. Easton
    p. cm.—(Oxford engineering science series; 50)
Includes bibliographical references and index.
ISBN 0-19-508545-0
1. Differentiable dynamical systems. 2. Iterative methods (Mathematics)
I. Title. II. Series.
QA614.8.E27 1998
515'.352—dc21   97-870

1 3 5 7 9 8 6 4 2

Printed in the United States of America
on acid-free paper

*To my parents,
Frances and Carson Easton*

Preface

One can view dynamics as the study of iteration processes. Iteration involves taking the output of a function and feeding it back as input. For example, suppose that you input a number and repeatedly press the cosine button on a pocket calculator. The number starting with the digits 0.739 eventually appears and remains on the display. This is explained by the technique of graphical analysis discussed in Chapter 1. Iteration processes are at the heart of many algorithms. They are used to numerically approximate solutions to ordinary and partial differential equations, and to numerically solve linear and nonlinear systems of equations.

Each iteration procedure generates a *discrete dynamical system*. The word "discrete" refers to fixing a time step and describing the state of a physical system at discrete instants of time that are integer multiples of the time step. However, the *phase space* or *state space* of the system is usually a connected subset of a Euclidean space. Mathematical models of continuously evolving physical systems can also be viewed as discrete dynamical systems by fixing a unit of time.

This book is meant to be a useful reference for mathematicians, engineers, and scientists who are studying dynamics. The reader is assumed to be familiar with linear algebra and calculus of several variables. My goal is to present the fundamental ideas clearly. The emphasis is on developing intuition and providing insight by using geometric methods. I have used some of the material presented here in teaching a first-year graduate course in dynamical systems for applied mathematicians. A one-semester course could be based on the first four chapters and Appendix A.

Illustrative examples of dynamical systems are presented in Chapter 1. Appendix A gives a review of the topology of metric spaces. The terminology and theorems presented in this appendix are used in a fundamental way throughout the text. Proofs are not included, since the material is presented in the cited references. It works well to ask students to prove for themselves and to present in class some of the basic topological results.

Chapter 2 gives basic definitions and results about the long-term behavior of the orbits of a dynamical system. It reflects a point of view or philosophy of dynamics which is strongly influenced by the work of Charles Conley. Rough orbits, or orbits with small errors, are used to define the chain recurrent sets to which all orbits eventually converge. The errors may arise as round-off errors in numerical simulations, or from approximations in the models used to describe physical processes, or from the effects of external influences.

The fundamental stable manifold and local structural stability theorems are presented in Chapter 3. The aim is to give insight into why these results are true. Proofs are given in a two-dimensional setting so that the reader can visualize the constructions. However, the methods easily generalize to higher dimensions. Hyperbolic invariant sets, homoclinic points, and homoclinic tangles are also discussed.

Isolated invariant sets and isolating blocks are defined in Chapter 4. One obtains information about the set of orbits trapped for all time inside a block by knowing the topology of the block and by knowing how the dynamical system maps the block across itself. A general symbolic dynamics is developed and used to give a powerful description of the orbit structures inside isolating blocks.

Chapter 5 develops what is called the Conley index in the context of discrete dynamics. The index associates quotient spaces and index maps with isolating blocks. An index homomorphism is defined on the homology groups of the index spaces, which mirrors at the algebraic level the way the block is mapped across itself. Algebraic conditions are derived which guarantee the existence of orbits that do not exit from a block.

Chapter 6 covers symplectic maps. Such maps are the discrete analog of Hamiltonian systems. Much of the material on symplectic structures and differential forms is prerequisite to a modern analysis of Hamiltonian dynamics. The treatments of generating functions and discrete variational principles are novel and, I hope, interesting.

The seventh and final chapter briefly introduces invariant measures. The Poincaré recurrence theorem is a fundamental result that has strong consequences for orbit behavior. Almost every orbit must return again and again close to where it started. Attracting sets do not exist, and questions concerning the stability of periodic orbits are very delicate.

The last three chapters require more from the reader's background than do the first four. However, I hope that much of the material is accessible to the nonexpert. Some knowledge of algebraic topology is necessary for a full reading of Chapter 5. Differential forms are used in Chapter 6, and some measure theory background is assumed in Chapter 7. Discussions are often open-ended and I hope that readers will want to investigate some of the topics at greater depth in the research literature and in their own research.

As a student at the University of Wisconsin, I received an excellent education from R. H. Bing, C. Conley, W. Rudin, A. Beck, S. Husseini, and others. I have learned a tremendous amount from the mathematicians who have attended the Midwest Dynamical Systems Seminar, including John Franks, Richard McGehee, Ken Meyer, Clark Robinson, and Bob Williams. I am most deeply indebted to Charles Conley, who guided my first research efforts, and who shared with me his understanding, insights, and friendship over many years. I want to thank Carson Chow, Beverly Diamond, Robert McLachlan, and Jim Meiss for many helpful suggestions and for criticisms of drafts of this book. The errors and obscurities which remain are my responsibility, and I would appreciate receiving suggestions for improvements.

Contents

1 Examples, 3
   A. Logistic Maps, 4
   B. Graphical Analysis, 5
   C. Hénon Maps, 9
   D. The Standard Map Family, 12
   E. Arnold's Circle Maps, 13
   F. Quadratic Maps, 14
   G. Duffing's Equation, 15
   H. Interesting Maps, 16
   I. Problems, 17
   J. Further Reading, 17

2 Dynamical Systems, 18
   A. Discrete and Continuous Dynamical Systems, 18
   B. Omega Limit Sets, 21
   C. Epsilon Chains, 25
   D. The Conley Decomposition Theorem, 28
   E. Directed Graphs, 30
   F. Local Analysis of Orbits, 31
   G. Summary, 33
   H. Problems, 33
   I. Further Reading, 34

3 Hyperbolic Fixed Points, 35
   A. Linearization, 35
   B. Stable and Unstable Manifolds, 39
   C. Shadowing and Structural Stability, 49
   D. The Hartman–Grobman Theorem, 51
   E. Smale's Horseshoe Map and Symbolic Dynamics, 54
   F. Hyperbolic Invariant Sets, 58
   G. Trellis Structure and Resonance Zones, 59

H. Topological Entropy, 70
I. Problems, 71
J. Further Reading, 72

4 Isolated Invariant Sets and Isolating Blocks, 73
A. Attracting Sets, 73
B. Isolated Invariant Sets and Isolating Blocks, 77
C. Constructing Isolating Blocks, 82
D. Basic Sets, 83
E. Symbolic Dynamics, 83
F. Filtrations of Isolated Invariant Sets, 85
G. Stacks of Isolating Blocks, 86
H. Calculating Directed Graphs, 88
I. Further Reading, 92

5 The Conley Index, 93
A. The Conley Index of an Isolating Block, 93
B. Continuation of Isolated Invariant Sets, 96
C. The Homology Conley Index, 98
D. References, 102

6 Symplectic Maps, 103
A. Linear Symplectic Maps, 103
B. Classical Mechanics, 107
C. Variational Principles, 110
D. Generating Functions, 113
E. Symplectic Integrators, 117
F. Separatrix Movement, 118
G. Normal Forms, 121
H. Problems, 125
I. Further Reading, 125

7 Invariant Measures, 126
A. Measure Spaces, 126
B. Invariant Measures, 127
C. Further Reading, 130

Appendix A   Metric Spaces, 131
A. Definitions, 131
B. The Hausdorff Metric, 134
C. Fractals, 135

Appendix B   Numerical Methods for Ordinary Differential Equations, 137

Appendix C   Tangent Bundles, Manifolds, and Differential Forms, 142

Appendix D Symplectic Manifolds, 147

Appendix E Algebraic Topology, 149

References, 152

Index, 155

Geometric Methods
for Discrete Dynamical Systems

1

Examples

A mathematical model of a physical system consists of a set which represents all "states" of the system together with a law which determines the time evolution of states. Part of a scientist's job is to identify the relevant set of states and to propose the law which governs their evolution. For example, states for Newton's model of the solar system consist of all positions and velocities of the sun and planets. The law of evolution is determined by a system of second-order differential equations. Newton's model does not include effects due to the spins, oblateness, magnetic fields of the sun and planets, or the effect of the solar wind. With most models, not every effect is taken into account, just the essential effects. Success of a model is judged by comparing its predictions with observations. Study of a successful model often provides new insight into the physical system being studied and it may suggest new experiments.

For Newton's model of the solar system, the differential equations governing the motion cannot be solved in terms of elementary functions. Attempts to solve these equations approximately using computers are successful over time periods of thousands of years, but they are not accurate over millions or hundreds of millions of years. Round-off errors and truncation errors due to the numerical method tend to grow over time and these eventually limit accuracy. Furthermore, deterministic chaos or instability in the dynamics may cause errors to grow exponentially in time. Attempts to model the solar system illustrate that even when the physical laws governing the evolution of a system are exactly known, and the initial state of the system is exactly known, there may be no practical way of predicting the state far in the future. This presents a new challenge to the modern scientific method which is just beginning to be appreciated and addressed.

In this book, the presentation is guided by these questions: What are the long term predictions of a model? Are predictions of the model "stable" in the sense that small changes in the initial state of the system or of the model itself produce small changes in the predictions? What can computer simulations tell us about the model?

We consider discrete dynamical systems as potential models for physical processes, but do not seek to model specific physical systems or to discuss the merits of specific models. Rather, we develop tools which may be used to help illuminate the predictions inherent in any given family of dynamical systems.

The examples in this chapter are meant to display a wide range of interesting phenomena. They are intended to excite your curiosity, and to pose challenges. They can be used to test whether theory is actually useful in practice. Aspects of these examples call for further research.

A. Logistic Maps

The most basic mathematical model of a physical process is obtained by iterating a function. The process of iteration involves taking the output of a function and feeding it back as input. We start by giving examples of the iteration process.

A well-studied family of nonlinear discrete dynamical systems is the *logistic map family*. Define

$$L: R^1 \to R^1; \quad x \to \alpha x(1-x)$$

The parameter α is a fixed real number. When α is between 0 and 4, the unit interval is folded and mapped into itself. A typical computer experiment is to input a value of α, and an initial "state" x_0 in the unit interval, and then to calculate the sequence of numbers

$$x_0, \quad x_1 = \alpha x_0(1-x_0), \quad x_2 = \alpha x_1(1-x_1), \quad \ldots$$

To explore the behaviour of such sequences or "orbits," repeat this experiment for a variety of initial states and parameter values.

Figure 1.1 shows the picture obtained as the parameter α steps between 2.8 and 4. For each fixed value of α the initial state $x = 0.5$ is chosen. The transient behaviour of the first one hundred iterates of x is not plotted. Instead, the figure consists of the second one hundred iterates of x plotted on the vertical axis above α. Thus for $\alpha = 2.8$, the hundred plotted points appear to coincide, whereas for $\alpha = 4$ the plotted points seem uniformly distributed on the unit interval. Our first goal is to explain to some extent Figure 1.1.

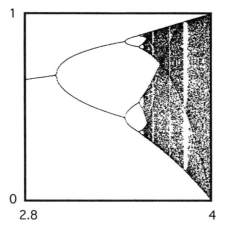

Figure 1.1. Logistic map orbits for a range of parameter values.

B. Graphical Analysis

The technique of graphical analysis allows one to picture the iteration process for a function defined on a subset of the line. First, draw the graph G of the function $y = f(x)$ in the x–y plane. Next, choose an initial state x and draw an arrow from the point (x, x) on the diagonal line $y = x$ to the point $(x, f(x))$ on the graph. Finally, draw a horizontal arrow from the point $(x, f(x))$ to the point $(f(x), f(x))$ back on the diagonal. Repeat this procedure to produce the sequence of points

$$(x, x), \quad (f(x), f(x)), \quad (f^2(x), f^2(x)), \quad \ldots$$

This sequence reveals the iterates of x pictured on the diagonal line $y = x$. The picture is used to gain rapid insight into the behaviours of the iterates of various points.

To see how this works, apply graphical analysis to the logistic map family. Where the graph crosses the line $y = x$, the function L has "fixed points" which are solutions of the equation

$$L(x) = x \quad \text{or} \quad \alpha x(1 - x) = x$$

The fixed points for L are $q = 0$ and $p = 1 - \alpha^{-1}$. The first and second derivatives of L are

$$L'(x) = \alpha(1 - 2x) \quad \text{and} \quad L''(x) = -2\alpha$$

Thus $L'(q) = \alpha$ and $L'(p) = 2 - \alpha$. The maximum value $\alpha/4$ of the function L occurs at $x = 1/2$.

Definition: To say a function is *smooth* means that its derivative exists and is continuous. A point p is an *attracting fixed point* for a smooth function f defined on the real line if

$$f(p) = p \quad \text{and} \quad |f'(p)| < 1$$

A point p is a *repelling fixed point* for a smooth function f defined on the real line if

$$f(p) = p \quad \text{and} \quad |f'(p)| > 1$$

The behavior of orbits which start near attracting fixed points is pictured in Figures 1.2a and 1.2b. The behavior of orbits which start near repelling fixed points is pictured in Figures 1.2c and 1.2d. The graph of a function near a fixed point is approximated by the tangent line at the fixed point, and the slope of this line determines whether orbits move toward or away from the fixed point.

LEMMA 1.B.1: If p is an attracting fixed point for a smooth function f, then there exists $h > 0$ such that for the interval $N = [p - h, \, p + h]$,

$$f(N) \subset N \quad \text{and} \quad \cap \, \{f^n(N) \colon n \geq 0\} = p$$

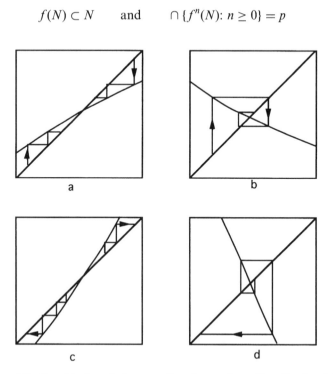

Figure 1.2. Graphical analysis near attracting and repelling fixed points.

Proof: Since f' is continuous, one can choose positive constants K and h so that

$$0 < K < 1 \quad \text{and} \quad |f'(x)| < K \quad \text{for all } x \in N$$

By the mean value theorem, for x belonging to N,

$$|f(x) - p| = |f(x) - f(p)| = |f'(c)||x - p| \quad \text{for some } c \in N$$

Thus $|f(x) - p| \leq K|x - p| \leq Kh$ and by induction,

$$|f^n(x) - p| \leq K^n|x - p| \leq K^n h.$$

Hence $f^n(N) \subset J_n = [p - K^n h, \; p + K^n h]$. The intervals J_n are nested and their intersection is $\{p\}$. Since $f^n(N) \subset J_n$, we have $\cap \{f^n(N) : n \geq 0\} = p$. ∎

LEMMA 1.B.2: If p is a repelling fixed point for a smooth function f, then there exists $h > 0$ such that for the interval $N = [p - h, \; p + h]$,

$$f(N) \supset N \quad \text{and for } x \in N - p, \; f^n(x) \notin N \quad \text{for some } n > 0$$

The proof is left as an exercise.

For the logistic map when α is between 0 and 1, $p = 1 - \alpha^{-1}$ is a repelling fixed point and 0 is an attracting fixed point. Graphical analysis applied to this case is indicated in Figure 1.3. There is a unique point $r > 1$ such that $L(r) = p$.

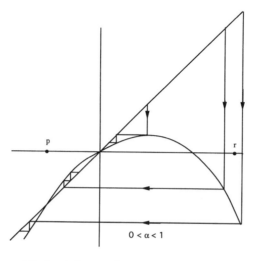

Figure 1.3. Logistic map for parameter range $0 < \alpha < 1$.

The orbits of points not in the interval $[p, r]$ converge to minus infinity, the orbit of 0 converges in one step to p, and the orbit of any point in the interval (p, r) converges to 0. Since we know the convergence behavior of all the orbits in this case, we can say that we understand the iteration process for these maps. Proofs of these assertions could be constructed using elementary calculus; however, the picture suggests the correct results.

When α is between 1 and 2, p is an attracting fixed point and 0 is a repelling fixed point. The graphical analysis is indicated in Figure 1.4. The orbits of points not in the interval $[0, 1]$ converge to minus infinity, the orbit of 1 converges in one step to 0, and the orbit of any point in the interval $(0, 1)$ converges to p.

When $\alpha = 2$, the fixed point p and the critical point $x = 1/2$ collide. The behavior of orbits is the same as described above.

When α is between 2 and 3, p is an attracting fixed point and 0 is a repelling fixed point. Graphical analysis applied to this case is pictured in Figure 1.5. The orbits of points not in the interval $[0, 1]$ converge to minus infinity, the orbit of 1 converges in one step to 0, and the orbit of any point in the interval $(0, 1)$ converges to p. This description is the same as the previous one. The only difference is that orbits converge to p not from one side of p but by alternating from one side to the other. The curve in Figure 1.1 above the α-interval $[2.8, 3]$ shows the position of the fixed point p as a function of α. The orbit of the initial state $x = 0.5$ evidently converges to p, and since the first one hundred iterates are not plotted, the plotted iterates almost coincide with p.

As α varies between 3 and 4, the picture becomes increasingly complicated. Both 0 and p are repelling fixed points. The first thing that happens is that p

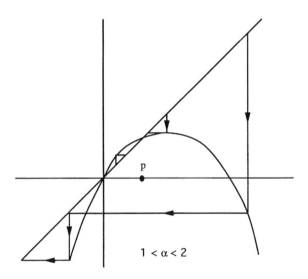

Figure 1.4. Logistic map for parameter range $1 < \alpha < 2$.

EXAMPLES

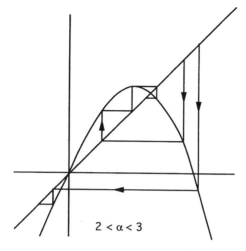

Figure 1.5. Logistic map for parameter range $2 < \alpha < 3$.

passes on its stability to an "attracting periodic point" z of period 2. This means that z is an attracting fixed point for L^2, but is not a fixed point for L. The orbit of z as a function of α appears above the α-interval [3, 3.4] as the parabolic-shaped curve which branches from the curve of fixed points above the α interval [2.8, 3]. As the parameter α continues to increase, z becomes "unstable" and passes its stability to an attracting periodic point of period 4. The branching continues into what is called a "period doubling cascade to chaos." An attracting orbit of period 3 emerges near the parameter value 3.84. This orbit loses stability as α increases, and another period doubling cascade to chaos occurs. From here on, things get even more complicated. See Figure 1.1.

When the parameter α is greater than 4, the orbits of points outside the interval [0, 1] converge to minus infinity. The unit interval is stretched and folded over itself, with the endpoints both mapping to the fixed point 0. The critical point 0.5 maps to $\alpha/4$, and an open interval U containing the critical point maps outside the interval [0, 1]. The set $[0, 1] - U$ consists of two closed intervals, each of which maps onto the interval [0,1]. One can show that the set of points whose iterates never leave the interval [0, 1] is homeomorphic to the standard (middle third) Cantor set. The iterates of all other points converge to minus infinity.

C. Hénon Maps

The Hénon family of maps is a two-parameter family of invertible maps of the plane. The original family was generated by composing three maps.

x-Compression $c: (x, y) \to (ax, y)$
Reflection $r: (x, y) \to (y, x)$
Shear $s: (x, y) \to (x, y + f(x)); \quad f(x) = b - x^2$

In order to preserve the orientation of the plane we will replace reflection with rotation

$$\rho: (x, y) \to (y, -x)$$

The resulting Hénon family depending on parameters a and b is the composition

$$H = s\rho c: (x, y) \to (ax, y) \to (y, -ax) \to (y, -ax + f(y))$$

The inverse of H is given by

$$H^{-1}: (x, y) \to (a^{-1}[-y + f(x)], x)$$

Definition: The *Jacobian matrix* of H is the matrix DH of first partial derivatives of H. The *Jacobian* of H is the determinant of the matrix DH where

$$DH(x, y) = \begin{pmatrix} 0 & 1 \\ -a & -2y \end{pmatrix}$$

In this case the Jacobian is constant and equal to a. Thus a rectangle having area R is mapped by H onto a set having area aR.

One may choose values for the parameters a and b, and an initial point (x_0, y_0), and generate an orbit or sequence of points in the plane

$$(x_0, y_0), \quad (x_1, y_1) = H(x_0, y_0), \quad (x_2, y_2) = H(x_1, y_1), \quad \ldots$$

Fixed points of H satisfy the equation $H(x, y) = (x, y)$. Therefore, $x = y$, and x solves the quadratic equation $x^2 + (a + 1)x - b = 0$.

The choice $a = 0.9$ and $b = 0$ gives interesting dynamics. For these parameter values the fixed points of H are $p = (-1.9, -1.9)$ and $q = (0, 0)$. Taylor's theorem provides a linear approximation to H near q:

$$A = DH(q) = \begin{pmatrix} 0 & 1 \\ -0.9 & 0 \end{pmatrix}$$

Iterating A instead of H, one generates the orbit $v, Av, A^2v, \ldots$ where v is an initial state treated as a column vector. Since $A^2v = 0.9v$, all orbits of A spiral toward the origin q. The same appears to be true for orbits of H which start near q. (It is only near q that A approximates H.) Figure 1.6 shows an orbit of H spiraling toward q with about 90° clockwise rotation.

Definition: An *attracting fixed point* for a smooth map f of R^2 is a point z such that $f(z) = z$, and the matrix of first partial derivatives of f at z, $Df(z)$, as all eigenvalues inside the unit circle. A *repelling fixed point* for a smooth map f of R^2 is a point z such that $f(z) = z$, and the matrix of first partial derivatives of f at z, $Df(z)$, has all eigenvalues outside the unit circle. A *saddle point* for a smooth map of f of R^2 is a point z such that $f(z) = z$, and the matrix of first partial derivatives of f at z, $Df(z)$, has one eigenvalue inside the unit circle and one eigenvalue outside the unit circle.

The fixed point q of H is attracting, but the fixed point p is a saddle point. It is neither attracting nor repelling, but it attracts and repels points on one-dimensional curves. The *stable manifold* of p consists of all points whose orbits converge to p. The *unstable manifold* of p consists of all points whose orbits with respect to H^{-1} converge to p. These manifolds are pictured as the curves in Figure 1.6. A central theorem in dynamics states that these attracted and repelled stable and unstable manifolds are indeed smooth curves.

Points where the stable and unstable manifolds of p intersect are called *homoclinic points*. Thus both forward and backward iterates of homoclinic points converge to p. The behavior of orbits which start near homoclinic points is complicated and interesting. This will be discussed in Chapter 3.

The linear approximation to H near p is determined by the matrix

$$B = DH(p) = \begin{pmatrix} 0 & 1 \\ -0.09 & 3.8 \end{pmatrix}$$

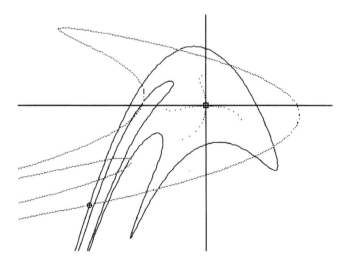

Figure 1.6. Henon map. Stable and unstable manifolds.

The eigenvalues of B are 0.2538, and 3.5462, and corresponding eigenvectors are (0.9693, 0.2460) and (0.2714, 0.9625). Consider the new coordinate system with coordinates (u, v) where $u = x + 1.9$ and $v = y + 1.9$. The origin of the u–v coordinate system is at the point p. Expressing H in terms of u and v gives $H(w) = Bw + Q(w)$, where $w = (u, v)$ is treated as a column vector and $Q(w)$ contains only quadratic terms.

The linear transformation B approximates H near p. The line spanned by the first eigenvector of B is the set of points whose iterates with respect to B converge to p. The line spanned by the second eigenvector of B is the set of points whose iterates with respect to B are repelled from p.

The curves in Figure 1.6 were generated by taking large numbers of evenly spaced points on four small line segments contained in the lines spanned by the eigenvectors of B. Points on the expanding line were marched forward by iterating H to generate the dark curve. Points on the contracting line were marched backward by iterating the inverse of H to generate the gray curve. In the following chapters, considerable effort will be devoted to explaining the dynamics pictured in Figure 1.6.

D. The Standard Map Family

The standard map family occurs as a very simplified model of an electron traveling in a cyclotron. The x-coordinate is proportional to the time at which the electron crosses a fixed plane perpendicular to the axis of the cyclotron, and the y-coordinate is proportional to the energy of the electron when it crosses this plane. The map is the composition of a nonlinear shear in the y-direction and a linear shear in the x-direction.

y-Shear $\quad s: (x, y) \to (x, y - f(x)); f(x) = a \sin(2\pi x)$

x-Shear $\quad \sigma: (x, y) \to (x + y, y)$

The *standard map family* depending on the parameter a is the composition $S = \sigma s$:

$$S: (x, y) \to (x, y - a \sin(2\pi x)) \to (x + y - a \sin(2\pi x), y - a \sin(2\pi x))$$

Since standard maps are periodic of period 1 in the x-coordinate, these maps may be thought of as maps of the cylinder obtained by identifying points in the plane whose x-coordinates differ by an integer. Standard maps are invertible and preserve area. Several orbits of the standard map for $a = 0.9$ are pictured in Figure 1.7. Orbits seem either to form "invariant curves" or to chaotically fill areas around the "islands" consisting of families of invariant curves.

The study of area-preserving maps such as the standard maps is a very active research area at present. The nature of the chaotic orbits is not under-

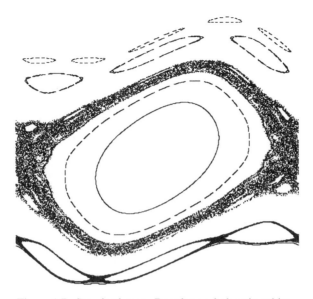

Figure 1.7. Standard map. Regular and chaotic orbits.

stood. The existence of invariant curves is predicted by famous results of Kolomogorov, Arnold, and Moser known as *KAM theory*. Invariant Cantor sets called Aubry–Mather sets are embedded in the chaotic bands. These sets tend to restrict passage of orbits across the bands.

E. Arnold's Circle Maps

A map of the line which depends periodically on x may be viewed as a map of the unit circle in the plane. Arnold's circle map is such an example. Define

$$A: R^1 \to R^1; \; x \to x + \omega + a \sin(2\pi x)$$

The map is periodic of period 1 and depends on two real parameters ω and a. The *rotation number* of a point x is defined to be

$$\rho(x, \omega, a) = \lim_{n \to \infty} [(A^n(x) - x)/n]$$

This limit exists when a is small and the limit is dependent of x. For larger values of the parameter a the map is no longer invertible and the limit may depend on x. Arnold studied how the rotation number depends on the parameters. It is important to note that the rotation number is an average that depends on whole orbits.

F. Quadratic Maps

A quadratic map $f: R^n \to R^n$ has the form $f(x) = (f_1(x), \ldots, f_n(x))$, where the component $f_j(x)$ is a quadratic polynomial in x which we represent as

$$f_j(x) = c_j + \sum_{k=1}^{n} a_j^k x_k + \sum_{k,l=1}^{n} \alpha_j^{kl} x_k x_l$$

There are $1 + n + n^2$ constants c_j, a_j^k, α_j^{kl} which appear in the formula for $f_j(x)$. Consequently, it takes 84 constants to specify a quadratic map of R^4. The task of exploring the family of quadratic maps of R^4 is essentially impossible. Even if just 10 values were chosen for each constant, the resulting collection of 10^{84} maps would be gigantic. We must be content with exploring interesting subfamilies of such maps, and we can ask what techniques will provide insight into the dynamics of any particular quadratic map.

An interesting subfamily of quadratic maps of R^4 is the family of symplectic quadratic maps. A *symplectic quadratic map* of R^4 is a quadratic map whose Jacobian matrix

$$Df(x) = \left(\frac{\partial f_i(x)}{\partial x_j} \right)$$

satisfies the equation $[Df(x)]^t J [Df(x)] = J$, where J is the 4×4 matrix

$$J = \begin{pmatrix} 0 & 0 & 1 & 0 \\ 0 & 0 & 0 & 1 \\ -1 & 0 & 0 & 0 \\ 0 & -1 & 0 & 0 \end{pmatrix}$$

It will be shown in Chapter 6 that any matrix that satisfies the equation $M^t J M = J$ has determinant 1. Therefore, symplectic maps preserve four-dimensional volume. The problem of establishing the stability of fixed points for such maps is very delicate. Since volume is preserved, such maps do not have attracting fixed points. It can be shown that if the Jacobian matrix of f at the fixed point has an eigenvalue that is not on the unit circle, then the fixed point is unstable. If all the eigenvalues are on the unit circle, then the fixed point is *elliptic*. The question of whether elliptic fixed points are stable is a difficult open question and is a focus of current research. Questions concerning the stability of the solar system (with Newton's law of gravitation), or of the stability of beams in particle accelerators and storage rings can be reduced to this question.

In a recent paper, Moser (1994) characterized symplectic quadratic maps. If one assumes that the map f has a fixed point, then coordinates can be chosen so that the fixed point is at the origin and the map has the form $f(x) = \sigma(Ax)$ where

$$A = \begin{pmatrix} 0 & C \\ C^t & 0 \end{pmatrix} \quad C = \begin{pmatrix} a & b \\ c & d \end{pmatrix} \quad C^t = \begin{pmatrix} a & c \\ b & d \end{pmatrix} \quad \det C = \pm 1$$

and σ is a "shear" map in four dimensions:

$$\sigma(x) = \left(x_1, x_2, x_3 + \frac{\partial V(x_1, x_2)}{\partial x_1}, x_4 + \frac{\partial V(x_1, x_2)}{\partial x_2} \right)$$

$$V(x_1, x_2) = \alpha x_1 + \beta x_2 + \gamma(x_1)^2 + (x_1)^3 + x_1(x_2)^2$$

Note the similarity with the Hénon maps. The dynamics of the above "six-parameter" family of symplectic quadratic maps is relatively unexplored. The origin is an elliptic fixed point whose stability characteristics are unknown. Surely new questions will be asked and new techniques developed by researchers trying to understand these maps.

G. Duffing's Equation

Duffing's equation is typical of the type of equation that occurs when a non-linear oscillator is subjected to a periodic applied force. The equation, written as a first-order system of differential equations, is

$$\dot{x} = y, \qquad \dot{y} + \alpha y - x + x^3 = \beta \cos(\gamma t)$$

When the parameters α, β are set to zero, the solutions of this equation are contained in the constant level curves of the function $H(x, y) = 0.5(y)^2 + 0.25(x)^4 - 0.5(x)^2$. This function measures the total energy of the oscillator, and is a conserved quantity. The level curve $H = 0$ has the shape of a figure-of-eight lying on its side with the crossing point at the origin (see Figure 1.8). A convenient way to study the behavior of solutions to this equation for positive values of α, β, and γ is to look at the associated stroboscopic or Poincaré map f. This map is a map of the plane defined by $f(x, y) = (u(2\pi/\gamma), v(2\pi/\gamma))$, where $(u(t), v(t))$ is the solution to the Duffing equation satisfying the initial condition $(u(0), v(0)) = (x, y)$. By iterating the map f one obtains the states of the oscillator at integer multiples of the period of the forcing function.

The behavior of solutions of the Duffing equation and the behavior of iterates of the map f are closely related. The map f may be numerically approximated by using one of the differential equation solvers discussed in Appendix B. However, the computations needed to iterate the map f once by this method are much more extensive that those needed to iterate a map given by a simple formula. To investigate the type of behaviour exhibited by solutions of Duffing's equation, it is much more efficient to study iterates of a

Figure 1.8. Duffing's equation. Constant energy curves.

map with a saddle point and having a figure-of-eight shaped homoclinic tangle. Such a map is given by the formula

$$G: (x, y) \to (y, -ax + by(1 - y^2))$$

with parameter values $a = b = 1$. This map may be viewed as being derived from a cubic map of the line using the ideas from the construction of the Hénon family of maps.

H. Interesting Maps

The *ABC* map is a volume-preserving map of a three-dimensional torus studied by Feingold, Kadanoff and Piro (Feingold, 1989). This map was used to model and study the motion of a drop of ink in a flowing incompressible fluid. The torus is the space $T^3 = R^3/Z^3$. The space is formed by identifying points in R^3 that differ by integers in each coordinate. Thus the points $(\sqrt{2}, \sqrt{2}, \sqrt{2})$ and $(\sqrt{2} + 1, \sqrt{2} - 3, \sqrt{2} + 2)$ are identified. The map $f = ABC$ is formed from the composition of three shear maps:

$C: (x, y, z) \to (x + c(y, z), y, z): c(y, z) = \alpha \sin(2\pi y) + \beta \cos(2\pi z)$

$B: (x, y, z) \to (x, y + b(x, z), z): b(x, z) = \gamma \sin(2\pi x) + \delta \cos(2\pi z)$

$A: (x, y, z) \to (x, y, z + a(x, y)): a(x, y) = \mu \sin(2\pi y) + \nu \cos(2\pi x)$

The composition preserves volume, and depends on the six parameters $\alpha, \beta, \gamma, \delta, \mu, \nu$. For some initial conditions, orbits seem to lie on the surface of two-dimensional tori, and for others, the orbits form a cloud with no aparent structure. Special choices of the parameters yield maps having real-valued functions called *integrals* which are constant on orbits. For example, the z-coordinate is preserved in the case where μ, ν are set to zero. It is interesting to study the orbits of maps which are close to the special "integrable" maps. A workstation with interactive software for visualization in three dimensions is quite valuable for such studies.

Gumowski and Mira (1980) study families of maps of the form $x_{n+1} = F(x_n) + G(Y_n)$, $y_{n+1} = -x_n + F(x_{n+1})$. Quadratic and cubic polynomials give some interesting choices of the functions F and G. The patterns formed by orbits can be very complicated and beautiful.

Lomeli and Meiss (1997) introduce and study a family of quadratic volume-preserving maps of the form

$$\begin{pmatrix} x' \\ y' \\ z' \end{pmatrix} = \begin{pmatrix} \alpha + \beta x + z + ax^2 + bxy + cy^2 \\ x \\ y \end{pmatrix}$$

Relatively little is known about the dynamics associated with this family, and how the stable and unstable manifolds of the two fixed points interact with each other. An interesting set of parameter values is the set $\alpha = 0$, $\tau = -0.3$, $a = c = 0.5$, $b = 0$. The stable and unstable manifolds of the fixed points intersect along an invariant curve that spirals from one fixed point to the other.

I. Problems

1. Prove Lemmma 1.B.2.
2. Prove versions of Lemmas 1.B.1 and 1.B.2 in two dimensions. First assume that the Jacobian matrix is diagonal. Explore numerically the case where the matrix has complex eigenvalues. Can you factor such a matrix into the product of a rotation matrix and a diagonal matrix?
3. Learn to use a computer to iterate maps and graphically display orbits. Investigate the Hénon family of maps and the family of maps defined in sections G and H.
4. Start your own collection of interesting maps and keep a list of references to these maps in the research literature.
5. Investigate the relationship between solutions of Duffing's equation and the stroboscopic map associated with this equation.
6. Prove the propositions stated in Appendix A.

J. Further Reading

A nice discussion on the logistic map family can be found in Devaney (1986). Many papers have been written about the Hénon maps. A selection of these papers include Hénon (1969), Benedicks and Carleson (1991), and Benedicks and Young (1993). Some papers about the standard map family are MacKay et al. (1984), Meiss (1992). Quadratic symplectic maps are discussed in Moser (1991).

2

Dynamical Systems

When scientists and engineers create models of the world, they use dynamical systems, and with these systems they predict how certain aspects of the world change. Their mathematical models allow them to see into the future and into the past.

The (planar) pendulum is a simple example to keep in mind as a physical system whose states and evolution can be modeled. The position of the pendulum bob is described by an angle and its velocity by the rate of change of the angle. These two quantities form the state of the pendulum system. As the pendulum moves, its state changes.

The "state" of a system collects exactly the information needed to determine its evolution. A "metric" is used to determine the distance between states. It is natural to assume that close initial states determine similar evolutions of the system at least over some appropriate period of time. The reader who is not familiar with the topology of metric spaces should read Appendix A in conjunction with this chapter since we consistently use metric spaces to model states of systems.

A. Discrete and Continuous Dynamical Systems

Definition: A *dynamical system* consists of a metric space X of all possible *states* of the system together with a function F which determines the time evolution of states. The set of all instants of time being considered is denoted by Γ. If a unit of time is chosen and the state is sampled after each passage of this unit, then $\Gamma = Z$ or Z^+ and time is said to be "discrete." Here Z denotes the set of integers, and Z^+ denotes the set of nonnegative integers. The set of real numbers is denoted by R^1, and the set of nonnegative real numbers is denoted by R^+. If $\Gamma = R^1$ or R^+, the time is said to be "continuous."

Formally, the function F is written $F: X \times \Gamma \to X$. The function F must be continuous and must satisfy the conditions

(1) $F(x, 0) = x$ and (2) $F(x, t + s) = F(F(x, t), s)$

Given that an initial state is x, the state of the system at time t is $F(x, t)$. Any continuous function satisfying conditions (1) and (2) is called a *flow*.

The *orbit* of x is the function $O_x: \Gamma \to X$ defined by $O_x(t) = F(x, t)$. The word "orbit" is also used to mean the set of states in the range of an orbit function.

There is an important correspondence between continuous-time dynamical systems and ordinary differential equations. Suppose that $X = R^n$, and that $F: R^n \times R^1 \to R^n$ is a smooth flow. One associates with this flow the function

$$G: R^n \to R^n; \quad G(x) = \dot{u}(0) \quad \text{with} \quad u(t) = F(x, t)$$

and the ordinary differential equation $\dot{x} = G(x)$. For example, for each $n \times n$ matrix A there is a flow on R^n defined by $F(x, t) = e^{tA}x$. The associated differential equation is $\dot{x} = Ax$. On the other hand, $u(t) = e^{tA}x$ is the unique solution of this differential equation which satisfies the initial condition $u(0) = x$. The content of the fundamental existence, uniqueness, and continuity theorems for ordinary differential equations can be summarized as follows:

THEOREM 2.A.1: If a function $G: R^n \to R^n$ is bounded and has continuous first partial derivatives, then for the differential equation $\dot{x} = G(x)$, there exists a unique differentiable flow $F: R^n \times R^1 \to R^n$ such that $u(t) = F(x, t)$ is the unique solution to the differential equation which satisfies the initial condition $u(0) = x$.

When time is discrete, the dynamical system is said to be a *discrete dynamical system*. By studying discrete dynamical systems one can avoid establishing Theorem 2.A.1 and one can begin immediately to study dynamics.

Definition: A *map* of a metric space X is a continuous function $f: X \to X$.
The discrete dynamical system *generated* by a map f is the function

$$F: X \times Z^+ \to X \quad \text{defined by} \quad F(x, n) = f^n(x)$$

where $f^0(x) \equiv x$ and $f^n(x) \equiv f(f^{n-1}(x))$.

Given a discrete dynamical system F, one can associate a map f defined by $f(x) = F(x, 1)$. From the definitions one can check that f generates F. Thus, maps f and dynamical systems of the type $F: X \times Z^+ \to X$ are paired with each other. The *orbit* of an initial state $x_0 \in X$ determined by a map f is the unique sequence $x_0, x_1, x_2, \ldots$, where $x_n = f^n(x_0)$.

The discrete dynamical system *generated* by the homeomorphism f is the function

$$F: X \times Z \to X \quad \text{defined by } F(x, n) = f^n(x)$$

where for $n < 0, f^n(x) \equiv (f^{-1})^{-n}(x)$. Each homeomorphism is paired with a discrete dynamical system and vice versa.

The goal when studying dynamics is to understand the behavior of orbits, since orbits predict the future states of the system. The simplest orbits are the periodic ones and much research has been devoted to finding them. In general it is important to study the long-term behavior of orbits. The question is: Given an initial state, what state or set of states is the system near after the passage of a long period of time? Analysis of a dynamical system is complete when the long-term behavior of all orbits is known.

An orbit is *periodic* if $x_n = x_0$ for some $n > 0$. The least such n is called the *period* of the orbit. A periodic point x of period 1 is called a *fixed point* since in this case $f(x) = x$.

An orbit is *eventually periodic* if

$$x_{m+n} = x_m \quad \text{for some } m \geq 0 \text{ and some } n > 0$$

An infinite sequence $\ldots, x_{-2}, x_{-1}, x_0$, is a *preorbit* of x_0 if

$$f(x_n) = x_{n+1} \quad \text{for each } n < 0.$$

Since f is not necessarily invertible, a point may not have a preorbit or it may have many preorbits. However, it has exactly one orbit. If f is onto then each point has at least one preorbit.

A *full orbit* is a bi-infinite sequence

$$\gamma: Z \to X \quad \text{such that } f(\gamma(j)) = \gamma(j+1) \text{ for each } j \in Z$$

Note that when f is a homeomorphism the terms orbit and full orbit have the same meaning, whereas for a map f an orbit is a sequence $\{f^n(x): n \geq 0\}$.

For $n \in Z^+$ and for a subset S of X, define

$$f^n(S) = \{f^n(x): x \in S\} \quad \text{and} \quad f^{-n}(S) = \{x: f^n(x) \in S\}$$

A *confining set* (or *forward invariant set*) for f is a subset T of X such that $f(T) \subset T$. Confining sets constrain the behavior of orbits, and finding such sets helps one to understand the dynamical system.

An *invariant set* for f is a subset I of X such that $f(I) = I$.

Propositions 2.A.2 and 2.A.3 will indicate that the closures of confining sets and invariant sets are also confining sets and invariant sets.

PROPOSITION 2.A.2: If $f(K) \subset K$, then $f(\text{cl}[K]) \subset \text{cl}[(K)]$.

Proof: Since f is continuous, $f(\text{cl}[K]) \subset \text{cl}[\,f(K)]$. Since $f(K) \subset K$, it follows that $\text{cl}[\,f(K)] \subset \text{cl}[K]$. ∎

PROPOSITION 2.A.3: The intersection of a family of confining sets is a confining set.

Proof: Let J_α for α in some index set A be confining sets. Let $J = \cap \{J_\alpha : \alpha \in A\}$. Choose a point x in J. Then x belongs to J_α. Since J_α is a confining set, $f(x)$ also belongs to J_α. Therefore $f(x)$ belongs to J, and hence $f(J) \subset J$. ∎

PROPOSITION 2.A.4: The closure of the union of a family of confining sets is a confining set.

Proof: The union of a family of confining sets is mapped inside itself. By Proposition 2.A.2, so is the closure of the union. ∎

PROPOSITION 2.A.5: If $f(J) = J$ and if $\text{cl}[J]$ is compact, then $f(\text{cl}[J]) = \text{cl}[J]$.

Proof: By Proposition 2.A.2, $f(\text{cl}[J]) \subset \text{cl}[f(J)]$. To show that $\text{cl}[J] \subset f(\text{cl}[J])$, choose a point x in the closure of J. Then there exists a sequence of points $\{x_n\}$ in J which converges to x. Since $f(J) = J$, there exists a sequence of points $\{y_n\}$ in J such that $f(y_n) = x_n$. Since $\text{cl}[J]$ is compact, there exists a subsequence $\{y_{n(k)}\}$ of the y's which converges to a point y in the closure of J. Since f is continuous,

$$f(y) = \lim_{k \to \infty} f(y_{n(k)}) = \lim_{k \to \infty} x_{n(k)} = x$$

Thus x belongs to $f(\text{cl}[J])$ and consequently $\text{cl}[J] \subset f(\text{cl}[J])$. ∎

B. Omega Limit Sets

Definition: The *omega limit set* $\omega(x, f)$ of a state x is the set of all points y such that a subsequence of the sequence $\{f^n(x) : n \geq 0\}$ converges to y.

The omega limit set is also denoted by $\omega(x)$ when the map is understood. A fundamental result concerning the long-term behavior of orbits is the following.

THEOREM 2.B.1: Suppose that the orbit of x is contained in a closed subset K of X. Then

(1) $\omega(x)$ is a closed subset of K and $f(\omega(x)) \subset \omega(x)$.

(2) If, in addition, K is compact, then $f(\omega(x)) = \omega(x)$ and the sequence $\{f^n(x): n \geq 0\}$ converges to $\omega(x)$.

Proof: Each point of $\omega(x)$ is a limit point of K. Since K is closed, $\omega(x)$ is contained in K. To show that $\omega(x)$ is closed, let q be a limit point of $\omega(x)$. Let $B(q, \varepsilon)$ be an epsilon neighborhood of q and let n be a positive integer. Since q is a limit point of $\omega(x)$, there exists a point p in $B(q, \varepsilon)$ which belongs to $\omega(x)$. Since p belongs to $\omega(x)$, there exists an integer $m > n$ such that $f^m(x)$ belongs to $B(q, \varepsilon)$. By choosing a sequence of epsilons converging to zero and an increasing sequence of m's one constructs a subsequence of the sequence $\{f^n(x)\}$ which converges to q. Hence q belongs to $\omega(x)$ and $\omega(x)$ is closed.

To show that $f(\omega(x)) \subset \omega(x)$, choose a point y belonging to $\omega(x)$ and a subsequence $f^{n(k)}(x)$ of points on the orbit of x converging to y. Since f is continuous, the sequence $f^{n(k)+1}(x)$ converges to $f(y)$. Hence $f(y)$ belongs to $\omega(x)$ and $f(\omega(x)) \subset \omega(x)$.

Suppose that K is compact. To show that $\omega(x) \subset f(\omega(x))$, choose a point y belonging to $\omega(x)$ and a subsequence $f^{n(k)}(x)$ of points on the orbit of x converging to y. Since K is compact, the sequence $f^{n(k)-1}(x)$ has a subsequence converging to some point z. It follows that z belongs to $\omega(x)$. Since f is continuous, and $\{f(f^{n(k)-1}(x)\}$ converges to y, it follows that $f(z) = y$. Thus $\omega(x) \subset f(\omega(x))$.

Finally, to show that the sequence $\{f^n(x)\}$ converges to $\omega(x)$, suppose not. Then there exists $r > 0$ and a subsequence $\{f^{n(j)}(x)\}$ which (by compactness of K) converges to a point p whose distance from $\omega(x)$ is a least r. However, from the definition of $\omega(x)$, p must belong to $\omega(x)$. This is a contradiction. ∎

The omega limit set is minimal in the following sense.

PROPOSITION 2.B.2: Suppose that K is a closed set such that $d(f^n(x), K) \to 0$ as $n \to \infty$. Then $\omega(x, f)$ is contained in K.

Proof: Suppose not. Then there exists a point y in $\omega(x, f)$ which does not belong to K. Since K is closed, there exists $\varepsilon > 0$ such that the distance from y to K is greater than ε. By hypothesis, N can be chosen such that for all $n > N$, $d(f^n(x), K) < \varepsilon/2$. Since y belongs to $\omega(x, f)$, there exists an integer $m > N$ such that $d(f^m(x), y) < \varepsilon/2$. There also exists a point k in K with $d(f^m(x), k) < \varepsilon/2$. By the triangle inequality, we have $d(y, K) < \varepsilon$, which is a contradiction. ∎

Unfortunately, the dependence of $\omega(x, f)$ on both x and f can be discontinuous.

Example: Suppose $X = [0, 1]$ and $f(x) = 2x(1 - x)$. Then

$$\omega(z, f) = \begin{cases} \frac{1}{2} & \text{if } x \neq 0, 1 \\ 0 & \text{if } x = 0, 1 \end{cases}$$

Thus $\omega(x, f)$ is not continuous in x.

Example: Suppose X is the set of real numbers and
$$f(x) = x - x(x-1)(x-0.5)^2$$
The fixed points for f are 0, 1, and 0.5. The omega limit set for x in the interval $(0, 0.5)$ is 0.5. Let $g(x) = f(x) + c$ where c is a small positive constant. The omega limit set with respect to g of any x in the interval $(0, 1)$ is not contained in the interval $[0, 1]$. Thus $\omega(x, f)$ is not continuous in f.

While omega limit sets clearly capture the long-term behavior of orbits, one would like to enclose them inside somewhat larger invariant sets which have better continuity properties with respect to the state x and the map f. This will be done in Chapter 4.

For Y a nonempty closed subset of X, define the *stable set* of Y to be the set
$$W^s(Y) = \{x \in X : d(f^n(x), f^n(Y)) \to 0 \text{ as } n \to \infty\}$$

Thus Theorem 2.B.1 can be restated as $x \in W^s(\omega(x))$.

PROPOSITION 2.B.3:
(1) If $A \subset B$ then $W^s(A) \subset W^s(B)$.
(2) If two invariant sets are a positive distance apart, then their stable sets are disjoint.

Proof: If not, then there is an orbit converging to both sets. This contradicts the definition of convergence. ∎

Definition: The *maximal invariant set* contained in a set Y is the set
$$\text{Inv}(Y) = \{x \in Y : \text{there exits a full orbit } \gamma : Z \to Y \text{ such that } \gamma(0) = x\}$$

PROPOSITION 2.B.4: The set $\text{Inv}(Y)$ is an invariant set. Further, if J is an invariant set which is contained in a set Y, then $J \subset \text{Inv}(Y)$.

The proof is left as an exercise.

The notion of omega limit set may be extended to include the omega limit set of a subset of initial states.

Definition: Let Y be a subset of X. A point x is an *omega limit point of Y* if given a neighborhood U of x and an integer m there exists an integer $n > m$ such that $f^n(Y)$ intersects U. The *omega limit set* $\omega(Y, f)$ of Y is defined to be the set of all omega limit points of Y.

THEOREM 2.B.5: If N is a compact confining set, then $\omega(N,f)$ is the maximal invariant set contained in N. Furthermore, $\omega(N,f) = \cap \, \{f^k(N): k \geq 0\}$.

Proof: Since each point of $\text{Inv}(N)$ has a preorbit in N, it follows from the definition of $\omega(N,f)$ that $\text{Inv}(N) \subset \omega(N,f)$.

To show that $\omega(N,f) \subset \text{Inv}(N)$, it is sufficient by Proposition 2.B.4 to show that $\omega(N,f)$ is an invariant set.

To show that $f(\omega(N,f)) \subset \omega(N,f)$, let y belong to $\omega(N,f)$. Let U be a neighborhood of $f(y)$ and let m be an integer. Since f is continuous, one can choose a neighborhood V of y such that $f(V)$ is contained in U. Since y belongs to $\omega(N,f)$, there exists $n > m$ and a point x in N such that $f^n(x)$ belongs to V. Then $f^{n+1}(x)$ belongs to U and therefore $f(y)$ belongs to $\omega(N,f)$.

To show that $\omega(N,f) \subset f(\omega(N,f))$, let x belong to $\omega(N,f)$. Then there exists a sequence of points y_k belonging to N and an increasing sequence of integers $n(k)$ such that $f^{n(k)}(y_k)$ converges to x. Let $z_k = f^{n(k)-1}(y_k)$. Since N is compact, the sequence $\{z_k\}$ has a subsequence which converges to a point z in N. Consequently, z belongs to $\omega(N,f)$. Since the sequence $f(z_k)$ converges to x and f is continuous, it follows that $f(z) = x$.

Points in the set $Z = \cap \, \{f^k(N): k \geq 0\}$ are omega limit points. It remains to show that $\omega(N,f) \subset Z$. It follows from the definition of $\omega(N,f)$, that for each $x \in \omega(N,f)$ there is a sequence of points $\{y_j\}$ contained in N and an increasing sequence of integers n_j such that the sequence $\{f^{n_j}(z_j)\}$ converges to x. Let $z_j = f^{n_j - k}(y_j)$. Since N is compact and confining, the sequence $\{z_j\}$ is contained in N and has a subsequence which converges to a point z in N. Since f is continuous and the sequence $\{f^k(z_j)\}$ converges to x, it follows that $f^k(z) = x$. Therefore, x is contained in $f^k(N)$; hence $\omega(N,f) = \cap \, \{f^k(N): k \geq 0\}$ ∎

Example: This example shows that Theorem 2.B.5 is false unless the confining set is assumed to be compact. Let X equal the nonnegative integers with the discrete topology. Inductively define the sequence $\{n_k\}$ by $n_1 = 2, n_k + 1 = n_k + k$. Define the map f on X by $f(j) = j - 1$ unless $j = 0$ or $j = 1$ or $j = n_k$ for some k. Define $f(0) = 0, f(1) = 0, f(n_k) = 1$ for $1 \leq k < \infty$. The map f has been constructed so the set $S = \cap \, \{f^k(X): k \geq 0\}$ is the set $\{0, 1\}$, and it is forward invariant but is not invariant (see Figure 2.1).

Figure 2.1. Example of a forward invariant set that is not an invariant set.

PROPOSITION 2.B.6. If f is a homeomorphism, then $\text{Inv}(Y) = \cap \{f^k(Y) : k \in Z\}$.

The proof is left as an exercise.

C. Epsilon Chains

A map f may not exactly represent all aspects of a physical system whose time evolution it models. The dynamics of nearby maps must also be considered. Computer simulations of the map f produce rough orbits whose relationships to true orbits must be investigated. Stable features of the orbit structure of f are those features which are shared by the orbit structures of maps "close" to f. Thus, rough orbits play an important role in discussing stability since an orbit for f may be a rough orbit for a nearby map g and vice versa. Rough orbits are orbits with errors. A mathematical analog of a rough orbit is an epsilon-chain. At each iteration of the function, an error of at most size epsilon is allowed.

Definition: An ε-*chain* is a finite sequence $y_0, y_1, y_2, \ldots$ of length at least 2 such that

$$d(y_{n+1}, f(y_n)) \leq \varepsilon \qquad \text{for each } n \geq 0$$

The following proposition states that epsilon-chains track orbits more and more closely as epsilon goes to zero.

PROPOSITION 2.C.1: Fix $n > 0$ and x. Given $\varepsilon > 0$ there exists $\delta > 0$ such that for any δ-chain $\{y_j\}$ which starts at x, $d(f^n(x), y_n) < \varepsilon$.

Proof: The proposition is evident for $n = 1$. We assume it is true for n, and show it is then true for $n + 1$. Using the continuity of f choose $\delta_1 < \varepsilon/2$ so that $d(f^n(x), z) < \delta_1$ implies that $d(f^{n+1}(x), f(z)) < \varepsilon/2$. By induction choose $\delta < \varepsilon/2$ so that $d(f^n(x), y_n) < \delta_1$ for any δ-chain $x, y_1, \ldots, y_n$. If y_{n+1} extends the δ-chain, then one has $d(f^{n+1}(x), y_{n+1}) \leq d(f^{n+1}(x), f(y_n)) + d(f(y_n), y_{n+1}) \leq \varepsilon$. ∎

Definition: A function f is *Lipschitz with constant k* it for all pairs of points x, y,

$$d(f(x), f(y)) \leq k d(x, y)$$

The next proposition gives an a priori bound on how closely an epsilon-chain tracks a true orbit. This is a useful error estimate when one solves differential equations numerically, and is a discrete version of the Gronwall inequality.

PROPOSITION 2.C.2: Suppose f is Lipschitz with constant k. Let $y_0, y_1, y_2, \ldots$ be an ε-chain. If x is an initial state, then

$$d(f^n(x), y_n) \leq k^n d(x, y_0) + \varepsilon(k^n - 1)/(k - 1) \quad \text{for } k \neq 1$$

and

$$d(f^n(x), y_n) \leq d(x, y_0) + \varepsilon n \quad \text{for } k = 1$$

Two lemmas are needed for the proof of this proposition.

LEMMA 2.C.3: Suppose that $f: R^1 \to R^1$ is a monotone increasing function. (This means that $f(x) \leq f(y)$ whenever $x \leq y$.) Suppose that $\{e_n\}$ is a sequence of numbers with $e_{n+1} \leq f(e_n)$ for all $n \geq 0$. Then $e_n \leq f^n(e_0)$ for all $n \geq 0$.

Proof: Assume that for a fixed n, $e_n \leq f^n(e_0)$. Then

$$e_{n+1} \leq f(e_n) \leq f^{n+1}(e_0)$$

By induction, we are done. ■

LEMMA 2.C.4: Suppose that $\{e_n\}$ is a sequence of numbers, and c, d are nonnegative constants such that $e_{n+1} \leq ce_n + d$ for all $n \geq 0$. Then

$$e_n \leq c^n e_0 + d(c^n - 1)/(c - 1) \quad \text{for } c \neq 1 \quad \text{and} \quad e_n \leq e_0 + dn \quad \text{for } c = 1$$

Proof: By the previous lemma, $e_n \leq f^n(e_0)$ for all $n \geq 0$, where $f(x) = cx + d$. Using induction, one computes that

$$f^n(e_0) = c^n e_0 + d(c^{n-1} + c^{n-2} + \cdots + 1) = c^n e_0 + d(c^n - 1)/(c - 1) \quad \blacksquare$$

Remark: For $r \geq 0$, $(0 + r) \leq e^r$ and hence $(1 + r)^n \leq e^{nr}$. Thus if $c = (1 + r)$, the previous lemma implies that

$$e_n \leq e^{nr} e_0 + dr^{-1}(e^{nr} - 1)$$

Proof of Proposition 2.C.2: Let $e_n = d(f^n(x), y_n)$. Then

$$e_{n+1} = d(f^{n+1}(x), y_{n+1}) \leq d(f^{n+1}(x), f(y_n)) + d(f(y_n), y_{n+1})$$

Hence, $e_{n+1} \leq d(f^n(x), y_n) + \varepsilon = ke_n + \varepsilon$.
The result now follows from Lemma 2.C.4. ■

Definition: Given a closed subset K of X and subsets A and B of X, the notation $\text{ch}(A, B, K, \varepsilon, f)$ will be used to denote the union of all ε-chains which start in A and end in B, and which are contained in K. As usual, f will be dropped from the notation unless other maps are being considered.

In the original discussion of chains by Charles Conley, the state space X was assumed to be compact. We have introduced the set K to localize the discussion of chains. The original definitions are recovered by setting $K = X$. When compactness is a necessary hypothesis, we will assume that K is compact.

The intersection of the sets $\text{ch}(A, B, K, \varepsilon, f)$ for all positive epsilon is denoted by $\text{ch}(A, B, K, f)$.

The *K-ε-chain recurrent* set of f is the set

$$\text{CR}(K, \varepsilon, f) = \{x \colon \text{ch}(x, x, K, \varepsilon, f) \text{ is nonempty}\}$$

The *K-Chain recurrent* set of f is the set

$$\text{CR}(K, f) = \cap \; \{\text{CR}(K, \varepsilon, f) \colon \varepsilon > 0\}$$

One may think of the chain recurrent set as the set of rough periodic points in K.

Example: Let X be a square in the plane. Let f be a map of X which is the identity map on the boundary of X and which fixes the x-coordinate and strictly decreases the y-coordinate of all points in the interior of the square. Then the chain recurrent set of f is all of X. However, the omega limit set of any point belongs to the boundary of X.

PROPOSITION 2.C.5: If $\text{ch}(x, x, K, \varepsilon) \cap \text{ch}(y, y, K, \varepsilon) \neq \varnothing$, then

$$\text{ch}(x, x, K, \varepsilon) = \text{ch}(y, y, K, \varepsilon)$$

Proof: Suppose that $z \in \text{ch}(x, x, K, \varepsilon) \cap \text{ch}(y, y, K, \varepsilon)$ and $w \in \text{ch}(y, y, K, \varepsilon)$. Then the sets $\text{ch}(x, z, K, \varepsilon)$, $\text{ch}(z, x, K, \varepsilon)$, $\text{ch}(y, z, K, \varepsilon)$, and $\text{ch}(z, y, K, \varepsilon)$ are nonempty. One can concatenate two chains whenever the last state of the first chain is equal to the initial state of the second chain. Therefore, one can chain from x to z, from z to y, from y to w, from w to y, from y to z, and from z to x. Hence, $w \in \text{ch}(x, x, K, \varepsilon)$. This shows that $\text{ch}(y, y, K, \varepsilon) \subset \text{ch}(x, x, K, \varepsilon)$. A similar argument shows that containment goes the other way. ∎

Definition: Define a relation on $\text{CR}(K, f)$ by setting $x \sim y$ if for each $\varepsilon > 0$ there exists a periodic ε-chain in K containing x and y. It follows from proposition 2.C.5 that this relation is an equivalence relation. The set $\text{ch}(x, x, K)$ is called the *K-chain equivalence class* of x. Equivalence classes are also called *K-basic sets*. These sets play a fundamental role in determining the long-term behavior of orbits.

D. The Conley Decomposition Theorem

The goal in this section is to group together states whose orbits have similar long-term behavior and thus to partition state space. In view of Theorem 2.B.1 it would be natural to partition X into equivalence classes where two states having the same omega limit set are equivalent. As we have already indicated, this partitioning is, in general, unstable and is difficult to describe. A somewhat coarser partition which groups "rough orbits" together seems to work better, and this partition is the essential feature of the Conley decomposition theorem (Theorem 2.D.4) developed below.

Lemma 2.D.1 and Propositions 2.D.2 and 2.D.3 establish properties of chains that will be used to prove the Conley decomposition theorem.

Let f be a map on a metric space X.

LEMMA 2.D.1: Given $\varepsilon > 0$ and given an orbit segment $x, f(x)$, and $f^2(x)$, there exist neighborhoods U, V, and W of $x, f(x)$, and $f^2(x)$, respectively, such that if $y_0 \in U, y_1 \in V$ and $y_2 \in W$ then y_0, y_1, y_2 is an ε-chain.

Proof: Using the continuity of f, choose U and V such that $f(U)$, V, and $f(V)$ are contained in $\varepsilon/2$ neighborhoods of the points $f(x), f(x)$, and $f^2(x)$, respectively. Choose W to be an $\varepsilon/2$ neighborhood of $f^2(x)$. One can verify that these neighborhoods have the desired property by using the triangle inequality for the metric on X. ■

PROPOSITION 2.D.2: Assume that K is compact. For any $x \in X$ whose orbit is contained in K and for any point $z \in \omega(x)$ the omega limit set of x is contained in the chain equivalence class of z. Thus $\omega(x) \subset \text{ch}(z, z, K)$.

Proof: Let $x_j = f^j(x)$ and let $z, w \in \omega(x)$. For any $\varepsilon > 0$ one can show that there exist positive integers n, m, r, such that the sequence $z, x_n, \ldots, n_{n+m}, w$, $x_{n+m+2}\ldots, x_{n+m+2+r}$, z is a periodic ε-chain in K. Since $f(z) \in \omega(x)$, there exists n such that $d(f(z), x_n) < \varepsilon$. Thus z, x_n forms an ε-chain. Since $w \in \omega(x)$ and $\omega(x)$ is invariant, there is a point $u \in \omega(x)$ with $f(u) = w$. By Lemma 2.D.1 we can choose neighborhoods U, V, W of the points $u, f(u), f^2(u)$, respectively, so that any three points chosen one from each neighborhood form an ε-chain. Since $u \in \omega(x)$, there exists a positive m such than $x_{n+m} \in U$ and $x_{n+m+2} \in W$. Hence $z, x_n, \ldots, x_{n+m}, w, x_{n+m+2}$ forms an ε-chain. Since $f(z) \in \omega(x)$, there exists $r + 1$ such that $d(z, x_{n+m+2+r+1}) < \varepsilon$. Thus the sequence $z, x_n, \ldots, x_{n+m}$, $w, x_{n+m+2}, x_{n+m+2+r}, z$, is the ε-chain. Therefore, $w \in \text{ch}(z, z, K)$ and $\omega(x) \subset \text{ch}(z, z, K)$. ■

PROPOSITION 2.D.3: If K is compact, and S is a subset of K, then $R = \text{ch}(S, S, K)$ is a compact invariant set contained in K.

Proof: To show that the set R is compact, it is sufficient to show that it is a closed subset of K. To do this, let x be a limit point of R (x belongs to K because K is closed). There is a point y in R such that $d(y, x) < \varepsilon/2$ and $d(f(y), f(x)) < \varepsilon/2$. Since y is in R there is an $\varepsilon/2$-chain $y_0, \ldots, y_r, \ldots, y_s$ which starts and ends in S, with $y_r = y$. Then $y_0, \ldots, x, \ldots, y_s$ is an ε-chain in K. Thus, x belongs to R so R is closed.

To show $f(R) \subset R$, let x be a point of R. It is sufficient to show that for any $\varepsilon > 0$, $f(x) \in \text{ch}(S, S, K, \varepsilon)$. Suppose that $f(x)$ does not belong to K. Then there is a neighborhood of x which maps outside K. For ε sufficiently small, any ε-chain through x must fall outside K. This contradicts the fact that x belongs to R. Thus $f(x)$ does belong to K. Choose $\delta < \varepsilon/3$ such that whenever $d(u, v) < \delta$ and $u, v, \in K$. $d(f(u), f(v)) < \varepsilon/3$ and $d(f^2(u), f^2(v)) < \varepsilon/3$. Choose a δ-chain $y_0, \ldots, y_n, \ldots, y_m$ in K with $y_0 \in S$ and $y_m \in S$ and $d(x, y_n) < \delta$. Then

$$d(f(x), y_{n+1}) \leq d(f(x), f(y_n)) + d(f(y_n), y_{n+1}) < \varepsilon$$

Also,

$$d(f^2(x), y_{n+2}) \leq d(f^2(x), f^2(y_n)) + d(f^2(y_n), f(y_{n+1})) + d(f(y_{n+1}), y_{n+2}) < \varepsilon$$

Hence the sequence $y_0, \ldots, y_n, f(x), y_{n+2}, \ldots, y_m$ is an ε-chain contained in K and $f(x) \in \text{ch}(S, S, K, \varepsilon)$.

To show that $R \subset f(R)$, let x be a point of R. For each positive integer n choose a $1/n$-chain $y_0, \ldots, y_{k(n)}, \ldots, y_m$ with $d(x, y_{k(n)}) < 1/n$. It follows that $d(x, f(y_{k(n)-1})) < 2/n$. Therefore, $f(y_{k(n)-1}) \to x$ as $n \to \infty$. Since K is compact there is a subsequence of the sequence $\{y_{k(n)-1}\}$ which converges to a point y in K. Because each point $\{y_{k(n)-1}\}$ belongs to $\text{ch}(S, S, K, 1/n)$ and R is closed, it follows that y belongs to R. Furthermore, $f(y) = x$, and thus x belongs to $f(R)$. Therefore, $R \subset f(R)$. ∎

THEOREM 2.D.4: Assume that K is a compact set. Then each K-basic set is an invariant set and K is the disjoint union of the stable sets of these sets, together with the set L of points in K whose orbits leave K.

Proof: Let $E = \text{ch}(x, x, K)$ be a basic set. E is an invariant set by Proposition 2.D.2. Any point x whose orbit is contained in K is in the stable set of its omega limit set. By Proposition 2.D.2, $\omega(x)$ is contained in a basic set. Any two basic sets are disjoint and compact. Thus they must be some positive distance from each other. But Proposition 2.B.3 their stable sets are disjoint. Therefore, K is partitioned as required. ∎

Exercise: Give an example where two chain equivalence classes are non-compact, zero distance apart, and have intersecting stable sets.

Each orbit which starts in a compact set K either exits from K or ultimately goes to the basic set which contains its omega limit set. The long-term behavior of orbits in K is determined by the K-chain recurrent set. The distinct homogeneous pieces of this set are the basic sets.

E. Directed Graphs

Definitions: The *directed graph of f in a compact set* K consists of a collection of vertices $V(E)$, one vertex for each K-basic set E, and a collection of directed edges. A directed edge goes from $V(E_1)$ to $V(E_2)$ if and only if for each $\varepsilon > 0$ there exists an ε-chain from the chain equivalence class E_1 to the chain equivalence class E_2. The directed graph expresses the gradient structure of the dynamical system inside K.

The *ε-directed graph of f in K* is obtained in the same fashion, where chain equivalence classes are replaced by ε-chain equivalence classes as vertices and edges are determined by ε-chains. This graph expresses and organizes the global information about the dynamics that one could hope to find using a computer simulation of the dynamics with ε-sized round-off errors.

Example: The system of differential equations $\dot{x} = x(1 - x)$, $\dot{y} = y(1 - y)$ has four equilibrium solutions at $(0, 0)$, $(0, 1)$, $(1, 0)$, and $(1, 1)$. Since the equations are decoupled, it is easy to see that solutions must behave as pictured in Figure 2.2. The arrows indicate that initial states which start on the grid of lines stay on these lines and move in time in the direction of the arrows. Let f denote the time one map associated with the flow ϕ determined by the differential equations. Then f has four fixed points, one repelling, two saddles, and one attracting.

The set $K = [-1, 2] \times [-1, 2]$ is compact and the K-basic sets of f are the four fixed points $E_1 = (0, 0)$, $E_2 = (1, 0)$, $E_3 = (0, 1)$, $E_4 = (1, 1)$. If L is the set

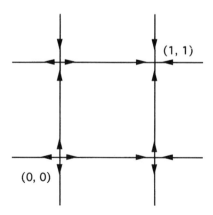

Figure 2.2. Four fixed points with connecting orbits.

of points which eventually leave K, then $K - L = [0, 2] \times [0, 2]$. The stable sets of these basic sets are the sets $W^s(E_1) = E_1$, $W^s(E_2) = (0, 2] \times 0$, $W^s(E_3) = 0 \times (0, 2]$, $W^s(E_4) = (0, 2] \times (0, 2]$. Thus, the Conley decomposition of K consists of five pieces. The directed graph associated with this decomposition is pictured in Figure 2.3.

One can tell a story to make these equations seem more interesting. The variable x may represent the fraction of the total carrying capacity of an environment occupied by a species x. The first differential equation is used to model the growth of the x-population which initially grows exponentially from a small initial population. As x gets near the carrying capacity, the negative interaction of the members of species x with each other decreases the growth rate and population growth slows down as x approaches 1. Species y behaves in the same way. Now suppose these populations interact; say that x inhibits the growth of y, but x is unaffected by y. One could change the model to the equations $\dot{x} = x(1 - x)$, $\dot{y} = y(1 - y) - (0.1)xy$. Now solutions are not as easy to picture. Add more species, and add a variety of interactions between them. One way to start to understand the dynamics is to start with uncoupled equations and to slowly turn on a coupling parameter. What happens? Just for fun, one could perioidically force the equations with a seasonal variation in environmental conditions, giving rise to the system of equations

$$\dot{x} = (1 - x) + y\sin(t), \qquad \dot{y} = y(1 - y) - (0.1)xy - x\sin(t)$$

Again, what happens?

Our understanding of dynamical systems is still rather rudimentary. Results in the subsequent chapters can be used to gain some insight into the behaviour of nonlinear dynamics, but there are no easy answers.

F. Local Analysis of Orbits

In many cases global analysis of a dynamical system is very difficult to achieve. However, a local analysis of orbits in some compact region of the state space

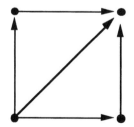

Figure 2.3. Directed graph.

may be possible. When studying the behavior of orbits in a compact set, it is useful to define functions which measure how long it takes for an orbit or a preorbit to leave the set. These functions are called exit time functions. Since time is discrete and the state space is usually connected, one expects these functions to be discontinuous. They are discontinuous, but in a predictable way, and their discontinuities reveal much about the orbit structure.

Definition: Let N be a compact set. Define *exit time functions* t^+ and t^- on N by

$$t^+(x) = \begin{cases} \infty & \text{if } f^n(x) \in N \text{ for all } n \geq 0 \\ \min\{j > 0 : f^j(x) \notin N\} & \text{if } f^n(x) \notin N \text{ for some } n > 0 \end{cases}$$

$$t^-(x) = \begin{cases} \max\{n \geq 0: \text{ for some } y \in N, \ f^j(y) \in N \text{ for } 0 \leq j \leq n \text{ and } f^n(y) = x\} \\ \infty \qquad \text{if the maximum does not exist} \end{cases}$$

N has *entry* and *exit sets* defined by $N_{\text{in}} = \{t^- = 0\}$ and $E = \{t^+ = 1\}$. Define the *exit threshold set* $e = \{x \in N : f(x) \in \partial N\}$. The set N has *exit time decompositions* into sets

$$E[i] = \{t^+ = i\} \quad \text{and} \quad E[i, j] = \{t^+ = i\} \cap \{t^- = j\}$$

Define $N[j] = \{t^+ \geq j\}$ and $N[j, k] = \{t^+ \geq j\} \cap \{t^- \geq k\}$.
Define $W^s(N, f) = E[\infty]$. The set $W^s(N, f)$ is the *stable set of N with respect to f*.

Remarks: When f is a homeomorphism, the subset of N consisting of all points with infinite exit time and infinite backward exit time is the maximal invariant set $\text{Inv}(N, f)$ contained in N. For large j and k the sets $N[j]$ and $N[j, k[$ approximate the sets $W^s(N, f)$ and $\text{Inv}(N, f)$, respectively.

Example: Let f be the linear map of the plane defined by $f(x, y) = (2x, y/2)$. Suppose that N is the unit square

$$N = \{(x, y) : 0 \leq x \leq 1, \ 0 \leq y \leq 1\}$$

Then

$$E[i] = \{(x, y) : 2^{-i} \leq x \leq 2^{-i+1}, \ 0 \leq y \leq 1\}$$

and

$$W^s(N, f) = \{(x, y) : x = 0, \ 0 \leq y \leq 1\}.$$

Example: Let f be the map of the set of complex numbers defined by $f(z) = z^2 + z$. The set of bounded orbits for this map is the famous Mandelbrot set (Devaney and Keen 1989). Note that $|f(z)| \geq |z|^2 - |z| = (|z| - 1)|z|$, where $|z|$ denotes the norm of the complex number z. By induction, we have $|f^n(z)| \geq (|z| - 1)^n |z|$. Thus, the orbit of a point z is unbounded whenever $|z| > 2$. Let D be the closed disk of radius 2 in the complex plane. Then the Mandelbrot set is the set $W^s(D,f)$. One can produce beautiful pictures in the following manner. Let N be the square in the complex plane consisting of all complex numbers whose real and imaginary parts belong to the interval $[-2, 2]$. Then $D \subset N$ and $W^s(D,f) = W^s(N,f)$. The square N is introduced since it is easily subdivided into a grid of smaller squares which we may identify with pixels on a computer screen. Suppose that the grid is 400 × 400 and that each pixel is colored according to the exit time assigned to the lower right corner of the pixel. Color the pixel black if the exit time exceeds a fixed number such as 1000. Then the set of black pixels approximates the Mandelbrot set.

A surprising amount of information about local dynamics can be obtained by studying exit time functions. In the next chapter, exit times will prove useful in the study of orbits in neighborhoods of fixed points. In Chapter 4 exit times will play an important role in the study of orbits near more complicated invariant sets.

G. Summary

A philosophy of dynamics emerges from the preceding results. Rough orbits or epsilon-chains play a central role. Chain recurrence expresses the idea of rough periodicity. The fate of orbits inside a compact set is determined by the Conley decomposition theorem. Essential steps in understanding the dynamics are to find the basic sets and the directed graphs. A local analysis of the behavior of orbits near basic sets will complete the picture.

H. Problems

1. Prove Proposition 2.B.4.
2. Prove Proposition 2.B.6.
3. Construct an example of a map of the plane having an orbit which does not converge to its (noncompact) omega limit set.
4. Write a computer program to generate and plot rough orbits of maps by adding random vectors of length less than a fixed magnitude to the output of the map. Test the sensitivity of some numerical differential equation solvers to truncation and round-off errors, by adding random "noise" of controlled magnitude to the numerical method. Try your

program on the Duffing equation or on the equation of a forced damped pendulum.
5. Give an example of a map of the plane having two chain equivalence classes which are noncompact, distance zero apart, and have intersecting stable sets.
6. Investigate the chain recurrent sets of maps in the logistic map family.
7. Let f be a map of a compact space X. A point in X is called a *wandering point* if there is an open set U containing x such that $f^n(U) \cap U = \varnothing$ for all positive integers n. The nonwandering set is the complement of the set of wandering points. Show that the nonwandering set is a subset of the chain recurrent set of f.
8. Formulate a dynamical system which models the interaction between several predator and prey species. Investigate whether the principle of competitive exclusion holds for your model. (The exclusion principle maintains that, of several predators on the same prey species, only one will prevail.)

I. Further Reading

There is a large and rapidly growing literature on dynamical systems. The following selection of books on the subject reflects my personal tastes, and is not comprehensive. For a general introduction see Devaney (1986), Devaney and Keen (1989), Arnol'd (1978) and Guckenheimer and Holmes (1983). More advanced treatments can be found in Ruelle (1989), Shub (1987), Guckenheimer et al. (1978), Bowen (1975), Conley (1978), Robinson (1995). Treatments of Hamiltonian dynamics can be found in Abraham and Marsden (1978), Arnol'd (1988), MacKay and Meiss (1987), Meyer and Hall (1991), Moser (1968), Moser (1973), Weinstein (1977).

3

Hyperbolic Fixed Points

A local analysis of orbits near fixed points of smooth maps is the primary focus of this chapter. The sets of points which converge to hyperbolic fixed points are called stable manifolds. They are shown to be Euclidian spaces nonlinearly imbedded in the given state space. Near a hyperbolic fixed point, the orbit structures of a map and its linear approximation are the same up to a homeomorphic change of coordinates.

A. Linearization

Definition: First partial derivatives of a function f from R^n to R^1 are defined by

$$D_j f(x) = \lim_{t \to 0} \left(\frac{1}{t}\right)[f(x + te_j) - f(x)]$$

where e_j is the jth standard basis vector for R^n. Thus, the first component of e_1 is 1 and all other components are 0. Second partial derivatives are defined as first partial derivates of the functions $D_j f$, and so on. A function f from R^n to R^1 is said to be *of class* C^r if all partial derivatives of f through order r are continuous. It is a theorem from advanced calculus that for a C^r function the order in which partial derivatives are taken does not change the resulting mixed partial derivative.

A function f from R^n to R^m is said to be *of class* C^r if each function $f_j(x)$ is C^r where $f(x) = (f_1(x), \ldots, f_m(x))$.

If a function f from R^n to R^m is of class C^1 then the *differential* (or *Jacobian Matrix*) $Df(x)$ is defined to be the $n \times m$ matrix

$$Df(x) = (D_j f_i(x))$$

The index j refers to the jth column of the matrix, and the index i refers to the ith row. From advanced calculus, the matrix $Df(x)$ is characterized as the unique matrix such that

$$\frac{|r(x, h)|}{|h|} \to 0 \quad \text{as } |h| \to 0$$

where $r(x, h) = f(x + h) - f(x) - Df(x)h$ and $|h| = [h_1^2 + \cdots + h_n^2]^{1/2}$.

Suppose that F and G are maps of metric spaces X and Y, respectively. F and G are said to be *topologically conjugate* if there exists a homeomorphism h from Y onto X such that

$$G = h^{-1}Fh$$

It follows that $h(G^n(y)) = F^n(h(y))$ for all $n \geq 0$. Thus h carries orbits of G onto orbits of F. Therefore images of fixed points and periodic points of G are fixed points and periodic points of F. We consider topologically conjugate maps as being equivalent. When X and Y are Euclidian spaces; the homeomorphism h may be viewed as a change of coordinates.

A nonsingular $n \times n$ matrix is *hyperbolic*, if none of its eigenvalues lies on the unit circle in the complex plane. A point p is a *hyperbolic fixed point* for a smooth map f on R^n if the matrix $Df(p)$ is hyperbolic. One can give a satisfactory analysis of the orbit structures of smooth maps near hyperbolic fixed points. However, in order to present the analysis in the most illuminating context, from now on we will concentrate on saddle points in the plane.

Suppose that f is a function from R^2 to R^2 of class C^r, where r is greater than or equal to 1. A point p is a *saddle point* of f if $Df(x)$ has real eigenvalues λ and μ such that

$$0 < |\mu| < 1 < |\lambda|$$

LEMMA 3.A.1: If M is a 2×2 matrix with distince real eigenvalues λ and μ, then there exists an invertible matrix L such that $L^{-1}ML$ is the diagonal matrix with diagonal entries λ and μ.

Proof: Choose nonzero eigenvectors v and w with $Mv = \lambda v$ and $Mw = \mu w$. To show that v and w are linearly independent, support that there exist scalars t and s such that $sv + tw = 0$. Then

$$0 = Msv + Mtw = s\lambda v + t\mu w = s\lambda v - s\mu v = s(\lambda - \mu)v$$

It follows that $s = 0$. Similarly, one shows that $t = 0$. Therefore v and w are linearly independent. Now form the matrix L with first column v and second column w. L is the desired matrix. ∎

HYPERBOLIC FIXED POINTS

Example: Consider the Hénon map discussed in Chapter 1. In the present notation,

$$f(x_1, x_2) = (x_2, -0.9x_1 - x_2^2)$$

There is a saddle point $p = (-1.9 - 1.9)$ for f with

$$Df(p) = \begin{pmatrix} 0 & 1 \\ -0.9 & 3.8 \end{pmatrix}$$

The eigenvalues for $Df(p)$ are approximately $\lambda = 3.5462$, $\mu = 0.2538$. The matrix L formed from the corresponding eigenvectors v and w is approximately

$$L = \begin{pmatrix} 0.2714 & 0.9693 \\ 0.9625 & 0.2460 \end{pmatrix}$$

The matrix L defines a linear change of coordinates so that the resulting map $L^{-1}ML$ has the x and y axes as eigenspaces.

The local analysis of a C^1 map f near a saddle point p begins with the following linearization procedure.

Linearization procedure

(a) Translate the origin to the point p by defining a translation

$$T: R^2 \to R^2; x \to x + p$$

(b) Calculate the matrix L such that

$$L^{-1}Df(p)L = \begin{pmatrix} \lambda & 0 \\ 0 & \mu \end{pmatrix}$$

(c) Define a new map g which is topologically conjugate to f by setting

$$g = L^{-1}T^{-1}fTL$$

(d) The map g is conjugate of f via the map TL. The origin is a fixed point for g and

$$Dg(0) = A = \begin{pmatrix} \lambda & 0 \\ 0 & \mu \end{pmatrix}$$

(e) From the approximation property of $Dg(0)$ we have

$$g(x) = Ax + e(x) \quad \text{with} \quad \frac{|e(x)|}{|x|} \to 0 \text{ as } |x| \to 0$$

(f) Since g is C^1, and $Dg(0) = A$, $|De(x)| \to 0$ as $|x| \to 0$.

A further useful step for the local analysis of the map f near the saddle point p is to magnify a neighborhood of p. This is accomplished by the following magnification procedure.

Magnification procedure

Define a linear contraction $C: R^2 \to R^2$; $x \to cx$ for $0 < c < 1$.

The map $F = C^{-1}L^{-1}T^{-1}fTLC$ is topologically conjugate to f via the map TLC. A map such as the map TLC is called an *affine map* when it is a composition of linear maps and translation maps. Affine maps transform straight lines to straight lines.

We have $F(0) = 0$, and $DF(0) = Dg(0)$. Furthermore, $F(x) = Ax + e(x; c)$, where $e(x; c) = c^{-1}e(cx)$.

Definition: The *box-norm* on R^2 is defined by $||x|| = \max\{|x_1|, |x_2|\}$. For a 2×2 matrix M with entries M_{ij}, the associated *matrix norm* is $||M|| = \max\{||Mx||: ||x|| = 1\}$. As an exercise, one can show using this definition that $||M||$ is calculated from the entries in M to be

$$||M|| = \max\{|M_{11}| + |M_{12}|, |M_{21}| + |M_{22}|\}$$

Define $N = \{x \in R^2 : ||x|| \leq 1\}$. The set N is a square and this is why the norm used to define it is called the box norm. A nice theorem from linear algebra states that all norms on a finite-dimensional Euclidian space are equivalent. In this case there are positive constants $k = 1/\sqrt{2}$ and $K = 1$ relating the box norm and the standard Euclidian norm: $k||x|| \leq |x| \leq K||x||$. The box norm will be used in the proof of the stable manifold theorems in the next section.

PROPOSITION 3.A.2: Given $\varepsilon > 0$ there exists $c > 0$ such that for any point x belonging to N,

(1) $||e(x; c)|| \leq \varepsilon ||x||$,
(2) $||De(x; c)|| \leq \varepsilon$.

Proof: Choose $c_1 > 0$ such that $||x|| \leq c_1$ implies that $||e(x)|| \leq \varepsilon ||x||$, where $e(x) = g(x) - Ax$. For x a point of N,

$$||e(x; c_1)|| = ||c_1^{-1}e(c_1 x)|| \leq c_1^{-1}\varepsilon ||c_1 x|| \leq \varepsilon ||x||$$

Since g is C^1, $||Dg(x) - A|| \to 0$ as $||x|| \to 0$. Thus, one can choose $c < c_1$ such that

$$||x|| \leq c \text{ implies } ||Dg(x) - A|| \leq \varepsilon$$

Note that $De(x; c) = De(cx)$. Therefore,

$$||De(x; c)|| = ||Dg(cx) - A|| \leq \varepsilon \qquad \blacksquare$$

Conclusion: By using the linearization and magnification procedures, we have reduced the study of saddle points to the analysis of maps of the form $f(x) = Ax + e(x)$, where the box-norm of $e(x)$ and the associated matrix norm of $De(x)$ are small uniformly on the square N.

B. Stable and Unstable Manifolds

Throughout this section the map f is assumed to be a continuous function from R^2 to R^2 with $f(x) = Ax + e(x)$ where $0 \leq |\mu| < 1 < |\lambda|$, and A is the diagonal matrix

$$A = \begin{pmatrix} \lambda & 0 \\ 0 & \mu \end{pmatrix}$$

Our goal is to show that the set of orbits of f trapped in the square $N = \{x \in R^2 : ||x|| \leq 1\}$ has the same topological structure as the set of orbits of the linear map A which are trapped in the square provided that the nonlinear term $e(x)$ is small.

Exit times are defined in section F of Chapter 2. The exit time decomposition of N with respect to the linear map A is pictured in Figure 3.1. The local stable set of N with respect to the linear map A is the dark vertical line segment passing through the origin.

Stable manifold theorems give information about the set $W^s(N, f)$. We discuss three theorems which give progressively stronger conclusions, depending on progressively stronger assumptions about the function f.

Definition: For $j = 1, 2$ let $\pi_j : R^2 \to R^1$, $\pi_j(x_1, x_2) = x_j$ be the projections onto the first and second coordinate axes, respectively.

THEOREM 3.B.1 (The Topological Stable Manifold Theorem): Let δ be a positive constant such that $\delta < 1$, $\delta < 1 - |\mu|$, $\delta < |\lambda| - 1$. Suppose that $||e(x)|| \leq \delta$ for all $x \in N$. Then for each $t \in [-1, 1]$ there exists $s \in [-1, 1]$ such that $(s, t) \in W^s(N, f)$.

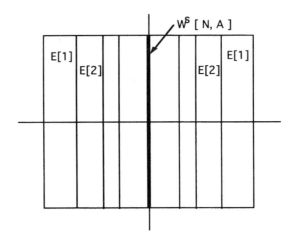

Figure 3.1. Exit time decomposition of N.

Proof: Assume that $\lambda > 1$. (The case $\lambda < -1$ is similar.) Let $x \in N$. If $x_1 = 1$, then $\pi_1 f(x) = \lambda + e_1(x) \geq \lambda - \delta > 1$ and hence $f(x) \notin N$. In the case where $x_1 = -1$, we have $\pi_1 f(x) < -1$. Thus, the right vertical side of the square N maps to the right of N and the left vertical side of the square N maps to the left of N.

For any point $x \in N$ we have $|\pi_2 f(x)| = |\mu x_2 + e_2(x)| < \mu + \delta < 1$. Thus N maps into the open horizontal strip $S = \{x \in R^2 : -1 < x_2 < 1\}$. Therefore, the orbits of points which exit from the square N exit either to the right or to the left of the square, and, as we will show, nearby points have the same exit behavior. Define open sets $R = (1, \infty) \times (-1, 1)$ and $L = (-\infty, -1) \times (-1, 1)$. Define sets $R^* = \{x \in N : F^{t^+(x)}(x) \in R$ and $L^* = \{x \in N : F^{t^+(x)}(x) \in L\}$.

Let $x \in R^*$ and consider the set of points $\{f^j(x) : 1 \leq j < t^+(x)\}$ on the orbit of x. If this set is contained in the interior of N, or is empty, then since f and its iterates are continuous, there exists a positive number ρ such that, if $\|x - u\| < \rho$, then $f^j(u) \in N$ for $1 \leq j < t^+(x)$ and $f^{t^+(x)}(u) \in R$. Therefore, all points of N within distance ρ of x belong to R^*.

Now suppose that $x \in N$ and for some j, $f^j(x)$ belongs to the boundary of N; then $j = t^+(x) - 1$ and $f^j(x)$ belongs to the right-hand vertical side of N. (The proof of this assertion involves eliminating all other possibilities and is left as an exercixe.) We have shown that the right vertical side of N maps into R. Hence all points close to this side do also. Thus, as before, there exists a positive number ρ such that if $\|x - u\| < \rho$, then $f^j(u) \in N$ for $1 \leq j < t^+(x)$ and $f^{t^+(x)}(u) \in R$. Therefore, R^* is open relative to N.

By the same argument, the set L^* is open relative to N. Since N is connected, it is not the union of these two disjoint open sets. Therefore the set $W^s(N, f)$ is not empty. Further, the horizontal line segment Γ which crosses the square N at "height" t is also a connected set which intersects both R^* and L^*.

HYPERBOLIC FIXED POINTS 41

Therefore, Γ must also intersect the set $W^s(N,f)$ at some point having coordinates (s, t). ∎

Theorems 3.B.2 and 3.B.3 stated below will be proven after several lemmas are established. By the linearization and magnification procedures we have shown that any C^1 map g of the plane with a saddle point p is topologically conjugate via an affine conjugacy to a map f satisfying the hypothesis of Theorem 3.B.3. Lemma 3.B.4 establishes that such a map also satisfies the hypothesis of Theorem 3.B.2. The box N may be thought of as the image under the conjugacy of a small neighborhood of the original saddle point p. Thus, these theorems describe the behavior of orbits near saddle points.

THEOREM 3.B.2 (The Litschitz Stable Manifold Theorem): Suppose that f is a Lipschitz function from R^2 to R^2. Given a positive constant c, choose and $\delta > 0$ so that

$$[c|\mu| + \delta(1 + c)] \le c[|\lambda| = \delta(1 + c)]$$

$$|\mu| + \delta(1 + c^{-1}) < 1 \quad \text{and} \quad |\lambda| - \delta(1 + c) > 1$$

Suppose that $\|e(x) - e(y)\| \le \delta \|x - y\|$ for all $x, y \in N$. Then there exists a Litschitz function $\gamma: [-1, 1] \to [-1, 1]$ with constant $1/c$ such that $\gamma(0) = 0$ and $W^s(N,f) = \{(\gamma(t), t) : t \in [-1, 1]\}$. In addition, for each point p in $W^s(N,f)$, the sequence $f^n(p) \to 0$ as $n \to \infty$.

THEOREM 3.B.3 (The Differentiable Stable Manifold Theorem): Suppose that f is a function from R^2 to R^2 of class C^r with $r \ge 1$ which satisfies the hypothesis of Theorem 3.B.2. Suppose that $\|De(x)\| \le \delta$ for all $x \in N$. Then there exists a C^r function $\gamma: [-1, 1] \to [-1, 1]$ such that $\gamma(0) = 0$, $\gamma'(0) = 0$ and

$$W^s(N,f) = \{(\gamma(t), t) : t \in [-1, 1]\}$$

Furthermore, for each point p in $W^s(N,f)$, the sequence $f^n(p) \to 0$ as $n \to \infty$.

LEMMA 3.B.4: Suppose that $\|De(x)\| \le \varepsilon$ on N. Then

$$\|e(x) - e(y)\| \le \varepsilon \|x - y\| \quad \text{for all } x, y \in N$$

Proof: Let $\gamma(t) = e_1(x + t(y - x))$. By the mean value theorem,

$$e_1(x) - e_1(y) = \gamma(1) - \gamma(0) = \gamma'(s) \quad \text{for some } s \in (0, 1)$$

By the chain rule, $\gamma'(s) = D_1 e_1(\gamma(s))(x_1 - y_1) + D_2 e_1(\gamma(s))(x_2 - y_2)$. Thus,

$$|e_1(x) - e_1(y)| \leq |D_1 e_1 \gamma(s)||x_1 - y_1| + |D_2 e_1(\gamma(s))||x_2 - y_2|$$

The box matrix norm associated with the box-norm that we are using gives

$$||De(x)|| = \max\{|D_1 e_1(x)| + |D_2 e_1(x)|, |D_1 e_2(x)| + |D_2 e_2(x)|\}$$

Hence, $|e_1(x) - e_1(y)| \leq \varepsilon ||x - y||$ and similarly $|e_2(x) - e_2(y)| \leq \varepsilon ||x - y||$. Therefore, $||e(x) - e(y)|| \leq \varepsilon ||x - y||$. ∎

Definition: Distinct points p and q which satisfy the condition $|p_2 - q_2| \leq c|p_1 - q_1|$ are said to form a c-*horizontal pair*. Points p and q which satisfy the condition $|p_2 - q_2| \geq c|p_1 - q_1|$ are said to form a c-*vertical pair*. A c-*horizontal slice* of N is defined to be a set Γ such that any pair of points in Γ is a c-horizontal pair and $\pi_1(\Gamma) = \pi_1(N)$.

A sketch of the proof of Theorem 3.B.2 will motivate the lemmas to follow. Lemma 3.B.5 shows that the image of a c-horizontal pair of points is a c-horizontal pair, and the horizontal distance between the image points increases. Lemma 3.B.6 shows that vertical distances between c-vertical pairs of points contract. Lemmas 3.B.5 and 3.B.6 are used to establish Proposition 3.B.7, which states that if Γ is a c-horizontal slice of N, then $f(\Gamma) \cap N$ is a c-horizontal slice of N.

The proof of Theorem 3.B.2 goes like this: Take any horizontal slice Γ of N. By Proposition 3.B.7 the set $f(\Gamma \cap N[2])$ is also a horizontal slice of N. Using induction, one shows that the set $f^j(\Gamma \cap N[j+1])$ is a horizontal slice of N. Thus the sets $\Gamma \cap N[j]$ are nonempty with $(\Gamma \cap N[j]) \supset (\Gamma \cap N[j+1])$ for each $j \geq 1$. They are also closed and hence their intersection is nonempty and is contained in $W^s(N, f)$. To show that this intersection contains exactly one point, one argues that if not, then there is a horizontal pair of points both of whose orbits remain in N. But Lemma 3.B.5 shows that the horizontal distance between successive iterates of these points must grow by a constant factor, and thus at some finite stage the horizontal distance between the iterates exceeds 2. Therefore, at most one of the iterates can be contained in N because all pairs of points in N have horizontal distance bounded by 2. This is a contradiction. Therefore, each horizontal line that intersects N intersects $W^s(N, f)$ in exactly one point. It follows that a function γ can be defined with $W^s(N, f) = \{(\gamma(t), t): t \in [-1, 1]\}$. With a little bit of thought one realizes that to show that the function γ is Lipschitz it is sufficient to show that every horizontal slice of N passing through a point $(\gamma(t), t)$ intersects the set $W^s(N, f)$ exactly in that point. Since we have already done this, the proof is complete. Figure 3.2 illustrates the proof. See also the proof of Theorem 3.B.8 for more details.

HYPERBOLIC FIXED POINTS 43

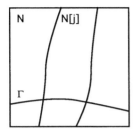
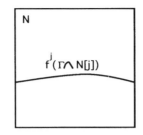
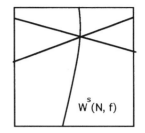

Figure 3.2. Horizontal slices. The right panel shows the cone of all horizontal slices passing through a point on the stable manifold.

LEMMA 3.B.5: Suppose that $||e(x) - e(y)|| \leq \delta ||x - y||$ for all $x, y \in N$. Suppose that $p, q \in N$ and $|p_2 - q_2| \leq c|p_1 - q_1|$. Let $P = f(p)$ and $Q = f(q)$. Then

(1) $|P_1 - Q_1| \geq [|\lambda| - \delta(c+1)]|p_1 - q_1|$
(2) $|P_2 - Q_2| \leq [c|\mu| + \delta(c+1)]|p_2 - q_2|$

Proof of (1)

$P - Q = A(p - q) + e(p) - e(q)$
$|P_1 - Q_1| = |\lambda(p_1 - q_1) + e_1(p) - e_1(q)| \geq |\lambda||(p_1 - q_1)| - ||e(p) - e(q)||$
$|P_1 - Q_1| \geq |\lambda||(p_1 - q_1)| - \delta||p - q|| \geq |\lambda||(p_1 - q_1)| - \delta(1 + c)|(p_1 - q_1)|$
$|P_1 - Q_1| \geq [|\lambda| - \delta(1 + c)]|(p_1 - q_1)|$

Proof of (2)

$|P_2 - Q_2| = |\mu(p_2 - q_2) + e_2(p) - e_2(q)| \leq |\mu||(p_2 - q_2)| + ||e(p) - e(q)||$
$|P_2 - Q_2| \leq |\mu||(p_2 - q_2)| + \delta||p - q|| \leq [c|\mu| + \delta(c+1)]|(p_1 - q_1)|$ ∎

LEMMA 3.B.6: Suppose that $||e(x) - e(y)|| \leq \delta ||x - y||$ for all $x, y \in N$. Suppose that p and q belong to N and let $P = f(p)$ and $Q = f(q)$. Suppose that $|p_2 - q_2| \geq c|p_1 - q_1|$ for $p, q \in N$. Then

$$|P_2 - Q_2| \leq [|\mu| + \delta(1 + c^{-1})]|p_2 - q_2|$$

Proof

$|P_2 - Q_2| = |\mu(p_2 - q_2) + e_2(p) - e_2(q)| \leq |\mu||p_2 - q_2| + ||e(p) - e(q)||$
$|P_2 - Q_2| \leq |\mu||(p_2 - q_2)| + \delta||p - q|| \leq [|\mu| + \delta(1 + c^{-1})]|(p_2 - q_2)|$ ∎

PROPOSITION 3.B.7: Suppose that, for all $x, y \in N$,

$$\|e(x)\| \leq \delta$$
$$\|e(x) - e(y)\| \leq \delta\|x - y\| \quad \text{for all } x, y \in N$$
$$|\mu| + \delta < 1 \quad \text{and} \quad |\lambda| - \delta > 1$$
$$[c|\mu| + \delta(1 + c)] \leq c[|\lambda| - \delta(1 + c)]$$

If Γ is a c-horizontal slice of N, then $f(\Gamma) \cap N$ is a c-horizontal slice of N.

Proof: Let p and q belong to Γ and let $P = f(p)$ and $Q = f(q)$. Then by Lemma 3.B.5,

$$|P_2 - Q_2| \leq [c|\mu| + \delta(c+1)]|p_1 - q_1| \leq c[|\lambda| - \delta(1+c)]|p_1 - q_1| \leq c|P_1 - Q_1|$$

To show that $f(\Gamma) \cap N$ a horizontal slice of N, it is sufficient by the inequality above to show that $\pi_1(f(\Gamma) \cap N) = \pi_1(N)$. Let a and b be the points in Γ with $\pi_1(a) = -1, \pi_1(b) = 1$. Γ is connected and therefore $f(\Gamma)$ is connected. Suppose that t is a point in the interval $[-1, 1]$. Define $U = \pi_1^{-1}((-\infty, t))$ and $V = \pi_1^{-1}((t, \infty))$. Since $|\lambda| - \delta > 1$ and $\pi_1 f(a) = -\lambda + e_1(a)$, it follows that $\pi_1(f(a)) < -1$ if $\lambda > 0$ and $\pi_1(f(a)) > 1$ if $\lambda < 0$. Similar results hold for $\pi_1(f(b))$. Therefore, the sets U and V both have nonempty intersection with $f(\Gamma)$ (the point $f(a)$ belongs to one set and the point $f(b)$ belongs to the other). Since $f(\Gamma)$ is connected, it must not be contained in the union of U and V. Hence there exists a point z in $f(\Gamma)$ with $\pi_1(z) = t$. Since $|\mu| + \delta < 1$ and $z = f(w)$ for some w in N, it follows that $|z_2| \leq |\mu| + |e_2(w)| \leq |\mu| + \delta$. Hence z belongs to N. Because t is an arbitaray point in the interval $[-1, 1]$, we have shown that $\pi_1(f(\Gamma) \cap N) = \pi_1(N)$. ∎

Definition: Suppose that $\{f_j\}$ is a sequence of functions from R^2 to R^2. Define F_n to be the composition $F_n(x) = f_n(\cdots(f_1(x))\cdots)$. Define

$$N[k, \{f_j\}] = \{x \in N : F_j(x) \in N \text{ for all } 1 \leq j \leq k\}, \text{ and } W^s(N, \{f_j\}) = N[\infty]$$

In the case where $f_j = f$ for all j, Theorem 3.B.8 below generalizes Theorem 3.B.2, the Lipschitz stable manifold theorem. The added generality of Theorem 3.B.8 is needed later for the proof of Theorem 3.B.3 and the Hartman–Grobeman theorem which is proven in section D and is further used to establish the shadowing property for hyperbolic invariant sets.

THEOREM 3.B.8

(1) Suppose that $\{f_j\}$ is a sequence of functions with $f_j(x) = Ax + E_j(x)$.
(2) Supose that $\|E_j(x)\| \leq \delta$ for all j.
(3) Suppose that $\|E_j(x) - E_j(y)\| \leq \delta\|x - y\|$ for all positive integers j.

(4) Suppose that positive constants c and δ have been chosen to satisfy the following conditions:

$$|\mu| + \delta(1 + c^{-1}) < 1, |\lambda| - \delta(1 + c) > 1$$
$$[c|\mu| + \delta(1 + c)] \leq c[|\lambda| - \delta(1 + c)]$$

(Thus $|\mu| + \delta < 1$ and $|\lambda| - \delta > 1$.)

Then there exists a Lipschitz function $\gamma: [-1, 1] \to [-1, 1]$, with Lipschitz constant c^{-1} such that

$$W^s(N, \{f_j\}) = \{(\gamma(t), t): -1 \leq t \leq 1\}$$

Furthermore,

$$\|F_n(p) - F_n(q)\| \to 0 \text{ as } n \to \infty \quad \text{for all } p, q \text{ in } W^s(N, \{f_j\})$$

Proof: Let Γ be the horizontal slice of N. By Proposition 3.B.7, $\pi_1 f(\Gamma)$ intersects N in a horizontal slice Γ_1. Similarly, $\pi_1 f(\Gamma_1)$ intersects N in a horizontal slice Γ_2. By induction, Γ intersects $N[k, \{f_j\}]$ for each k. Thus Γ intersects $W^s(N, \{f_j\})$.

To show that Γ intersects $W^s(N\{f_j\})$ in a unique point, let p and q be points in the intersection. By Lemma 3.B.5, $|\pi_1(f_1(p) - f_1(q))| \geq k|p_1 - q_1|$ where $k = |\lambda| - \delta(c + 1)$. By Proposition 3.B.7 the points $f_1(p)$ and $f_1(q)$ remain c-horizontal. This argument can be repeated to show that $|\pi_1(F_n(p) - F_n(q))| \geq k^n|p_1 - q_1|$. Since $k > 1$, this implies that $|p_1 - q_1| = 0$. Since p and q belong to Γ, their second coordinates must also be equal. Hence $p = q$.

Let $(\gamma(t), t)$ be the unique point where the slice $\Gamma = \{(s, t): -1 \leq s \leq 1\}$ intersects $W^s(N, \{f_j\})$. This implicity defines the function γ.

To show that γ is a Lipschitz function, let $p = (\gamma(t), t)$ and $q + (\gamma(s), s)$. Then we have shown that p and q cannot lie on a c-horizontal slice. Therefore, $|t - s| \geq c|\gamma(t) - (s)|$. Hence, the Lipschitz constant for γ is $1/c$.

Suppose that p and q belong to $W^s(N, \{f_j\})$. To show that $\|F_n(p) - F_n(q)\| \to 0$ as $n \to \infty$, use Lemma 3.B.6 repeatedly. The points $F_n(p)$ and $F_n(q)$ are never horizontally related. Thus by Lemma 3.B.6, $|\pi_2(F_n(p) - F_n(q))| \leq \kappa^n|p_2 - q_2|$, where π_2 denotes projection onto the second coordinate and $\kappa = |\mu| + \delta(1 + c^{-1})$. By hypothesis, $\kappa < 1$ and the vertical distance between $F_n(p)$ and $F_n(q)$ converges to zero, Since these points are not horizontally related, the horizontal distance between them also goes to zero. ∎

COROLLARY 3.B.9: Theorem 3.B.2 follows from Theorem 3.B.8 when $f_j = f$ for each j.

The remainder of this section is devoted to the proof of Theorem 3.B.3, and is concerned with the smoothness of the function γ. If the function γ is differentiable, then every vector tangent to the stable manifold $W^s(N, f)$ at the point p should belong to the set $W^s[V, \{A_j\}]$ defined below. Each vector in the sequence of vectors $\{Df^j(p)(v)\}$ should be a vertical vector tangent to the stable manifold. Conversely, we will show that every vector in $W^s[V, \{A_j\}]$ is a tangent vector. Thus, to find the tangent line to the graph of γ at a point p we first construct the set $W^s[V, \{A_j\}]$.

Definition: Let p belong to $W^s(N, f)$ and let $p_j = f^j(p)$. Define $A_j = Df(p_j)$. Define the set $V = \{v \in R^2 : c|\pi_1(v)| \leq |\pi_2(v)|\}$. The set V looks like a bow-tie rotated 90°. Define $V[n] = \{v \in V : A_{j-1}A_{j-2} \cdots A_0(v) \in V \text{ for } 1 \leq j \leq n\}$ and define $W^s[V, \{A_j\}] = V[\infty]$.

PROPOSITION 3.B.10: Assume that the hypothesis of Theorem 3.B.8 is satisfied with $f_j = f$ for each j, and assume that the function f is C^1. Then the set $W^s[V, \{A_j\}]$ is a straight line.

Proof: The sequence of maps $\{A_j\}$ satisfies the hypothesis of Theorem 3.B.8. Thus the set $W^s[N, \{A_j\}]$ is nonempty and is contained in the set $W^s[V, \{A_j\}]$. Since the maps A_j are linear, if $v \in W^s[V\{A_j\}]$, then $tv \in W^s[V, \{A_j\}]$ for all $t \in R^1$. To show that the set $W^s[V, \{A_j\}]$ is a straight line it is sufficient to show that it intersects the horizontal line of height 1 in a single point. Thus, consider two vectors v and w belonging to $W^s[V, \{A_j\}]$ with $\pi_2(v) = \pi_2(w) = 1$. The vector $v - w$ is a horiozntal vector and by Lemma 3.B.5 the horizontal distance between the images grows by the factor k. Thus for v and w to belong to $V[n]$, we must have $|\pi_1(v - w)| \leq 2ck^{-n}$. It follows that $v = w$. ∎

The next lemma applies to all differentiable functions, and relates the derivative to approximating sequences of points and vectors.

LEMMA 3.B.11: Suppose that f is a C^1 function and suppose that

$$z_n \to p \text{ and } v_n \to v \text{ where } v_n = \frac{(z_n - p)}{\|z_n - p\|}$$

Then

$$w_n \to Df(p)v \text{ where } w_n = [f(z_n) - f(p)]\|z_n - p\|$$

Proof: From the definition of Df,

$$\|w_n - Df(p)v_n\| \to 0 \text{ as } n \to \infty$$

since $v_n \to v$, $Df(p)v_n \to Df(p)v$. Hence $w_n \to Df(p)v$. ∎

HYPERBOLIC FIXED POINTS

PROPOSITION 3.B.12: If f is a C_1 function and the hypothesis of Theorem 3.B.3 is satisfied, then the function γ is differentiable. Further,

$$W^s(V, \{A_j\}) = \{(\gamma'(p)t, t) : t \in R^1\}$$

Proof: One needs to show that

$$\lim_{\tau \to t} \frac{\gamma(\tau) - \gamma(t)}{\tau - t}$$

exists. If not, then since the function γ is Lipshitz, there exist sequences $t_n \to t$ and $s_n \to t$ such that the limits

$$\lim_{n \to \infty} \frac{\gamma(t_n) - \gamma(t)}{t_n - t} \quad \text{and} \quad \lim_{n \to \infty} \frac{\gamma(s_n) - \gamma(t)}{s_n - t}$$

exist and are not equal. Define

$$z_n = (\gamma(t_n), t_n) \quad \text{and} \quad v_n = \left(\frac{\gamma(t_n) - \gamma(t)}{t_n - t}, 1\right)$$

$$q_n = (\gamma(s_n), s_n) \quad \text{and} \quad u_n = \left(\frac{\gamma(s_n) - \gamma(t)}{s_n - t}, 1\right)$$

Set

$$p = (\gamma(t), t) \quad v = \lim_{n \to \infty} v_n \quad \text{and} \quad u = \lim_{n \to \infty} u_n$$

Then $z_n \to p$, $v_n \to v$, $u_n \to u$ and $v \neq u$. We will show that $u, v \in W^s(V, \{A_j\})$. Then by Proposition 3.B.10, the horizontal line $\Gamma = \{(s, 1) : s \in R^1\}$ intersects the set $W^s(V, \{A_j\})$ in at most one point. Therefore, $u = v$. This is a contradiction. Thus, the limit exists and therefore γ is differentiable.

We finish the proof by showing that $u, v \in W^s(V, \{A_j\})$. The function f^n is a C^1 function, and Lemma 3.B.11 implies that

$$w_n \to Df^n(p)v \quad \text{where} \quad w_n = \frac{[f^n(z_n) - f^n(p)]}{\|z_n - p\|}$$

Since $\{z_n\}$ is a sequence of points in $W^s(N, f)$, the vectors w_n are "vertical" and thus belong to V. Since V is closed, it follows that $Df^n(p)v$ belongs to V. Since n is arbitrary, we conclude that $v \in W^s(N, \{A_j\})$. Similary, one shows that $u \in W^s(V, \{A_j\})$. ∎

To show that γ is C^1 we need to show that γ' is continuous.

PROPOSITION 3.B.13: γ' is continuous.

Proof: Let $p = (\gamma(t), t)$, $q = (\gamma(s), s)$. Given $\varepsilon > 0$ we must be able to choose $\delta > 0$ so that $|\gamma'(t) - \gamma'(s)| \leq \varepsilon$ when $|t - s| < \delta$. To do this, choose n such that $2k^{-n} \leq \varepsilon$. Set $y_0 = (\gamma'(t), 1)$ and $w_0 = (\gamma'(s), 1)$. Set $y_j = A_{j-1} \cdots A_1(y_0)$, set $w_j = B_{j-1} \cdots B_1(w_0)$, and set $z_j = B_{j-1} \cdots B_1(y_0)$. Note that y_j belongs to the interior of V. By continuity of f, Df, and γ, there exists $\delta > 0$ such that if $|t - s| < \delta$ then $z_j \in V$ for $1 \leq j \leq n$. The vector $y_0 - w_0$ is horizontal. Thus by Lemma 3.B.5 we have $k^n |\pi_1(w_0 - u_0)| \leq |\pi_1(w_n - u_n)| \leq 2$. Therefore, $|\gamma'(t) - \gamma'(s)| = |\pi_1(y_0 - w_0)| \leq 2k^{-n} \leq \varepsilon$. ∎

One way to prove that γ is C^r when the map f is C^r is to establish a general stable manifold theorem in R^d and then to use the "bootstrap" argument of Theorem 3.B.15. A very brief outline of this argument follows.

HYPOTHESIS: Suppose that $F: R^u \times R^s \to R^u \times R^s$ is a C^1 function with

$$F(x_1, x_2) = (Ax_1, Bx_2) + E(x)$$

where A and B are $u \times u$ and $s \times s$ matrices, respectively. Suppose that $\|\ \|_u$ and $\|\ \|_s$ are norms on R^u and R^s. Suppose that $\|Ax_1\|_u \geq \lambda \|x_1\|_u$ and $\|Bx_2\|_s \leq \mu \|x_2\|_s$ with $0 < \mu < 1 < \lambda$. Suppose that $\|\ \|$ is the "box" norm on $R^u \times R^s$ defined by $\|x\| = \max\{\|x_1\|_u, \|x_2\|_s\}$. Let

$$N = \{x : \|x\| \leq 1\}$$

$$N^u = \{x_1 \in R^u : \|x_1\|_u \leq 1\} \qquad N^s = \{x_2 \in R^s : \|x_2\|_s \leq 1\}$$

Suppose that $\|E(x)\| \leq \varepsilon$ and $\|DE(x)\| \leq \varepsilon$ when $x \in N$.

Definition: Let $\pi_u : R^u \times R^s \to R^u$, $\pi_u(x, y) = x$ and $\pi_s : R^u \times R^s \to R^s$, $\pi_s(x, y) = x$ be projections. Generalizing the previous definitions of horizontal pairs and slices, we say that distinct points p and q which satisfy the condition $|\pi_s(p - q)| \leq c |\pi_u(p - q)|$ form a *c-horizontal pair*. A *c-horizontal slice* of N is defined to be a set Γ such that any pair of points in Γ is a c-horizontal pair and $\pi_1(\Gamma) = \pi_u(N)$.

Definition: Since F is C^1 one may define the associated *tangent map*

$$TF : R^u \times R^s \times R^u \times R^s \to R^u \times R^s \times R^u \times R^s$$

by the formula

$$TF(x, y) = (F(x), DF(x)y)$$

One can generalize the previous arguments to prove the following theorem.

THEOREM 3.B.14: Suppose that F satisfies the preceeding hypothesis. If ε is sufficiently small, then there exists a C^1 function $\gamma: N^s \to N^u$ such that

(1) $W^s(N, F) = \{(\gamma(t), t): t \in N^s\}$
(2) $W^s(N \times N, TF) = \{(\gamma(t), t, D\gamma(t)v, v): t \in N^s \text{ and } v \in N^s\}$

THEOREM 3.B.15: Assume the hypothesis of Theorem 3.B.14, and in addition assume that F is C^r. Then there exists a C^r function $\gamma: N^s \to N^u$ such that the conclusions (1) and (2) of Theorem 3.B.14 are true.

Proof: The proof is by induction. For $r = 1$, the theorem is true by Theorem 3.B.14. Assume that it is true for $r = k - 1$. Then TF is C^{k-1}. One can show that $G = TF$ consists of a hyperbolic linear part plus a C^1 small nonlinear pertubation. By inducation, we have

$$W^s(N \times N, TF) = \{(\gamma(t), t, D\gamma(t)v, v): t \in N^s \text{ and } v \in N^s\}.$$

Furthermore, by induction the function $D\gamma(t)v$ must be C^{k-1}. Therefore γ is C^k. ∎

C. Shadowing and Structural Stability

The picture formed by orbits of a map is called its *phase portrait*. It is natural to ask how the phase portrait changes if the map is changed slightly. If the picture is topologically the same, then the map is *structurally stable*.

The set of all maps from X to X is denoted by $C(X)$. Suppose that Ψ is a subset of $C(X)$. The set Ψ can be considered as the set of possible models for a process. Further, suppose that Ψ is a metric space whose metric measures in some natural way how close two maps are to each other.

Definition: A map f belonging to Ψ is structurally stable if there is an open subset U of Ψ such that each map g belonging to U is topologically conjugate to f.

This is a very strong condition, and it is not easy to verify whether a given map is structurally stable. Further, the stability is measured relative to the space Ψ of maps being considered. If Ψ is replaced by a larger space or a different metric is placed on Ψ, then a structurally stable map may become unstable in this new context.

Example: Let X be a compact subset of R^n. Let Ψ be the set of C^1 maps from X to X. The C^1-metric on Ψ is defined by the formula

$$d(f, g) = \sup\{|f(x) - g(x)|: x \in X\}$$
$$+ \sup\{|D_j f(x) - D_j g(x)|: x \in X, 1 \leq j \leq n\}$$

Example: Let $X = R^1$, and let Ψ be the family of logistic maps. Define a metric by $d(f, g) = |\alpha - \beta|$ when $f(x) = \alpha x(1 - x)$, and $g(x) = \beta x(1 - x)$.

Exercise: Find a logistic map that is not structurally stable in Ψ. Justify your answer.

Exercise: Find a logistic map that is structurally stable in Ψ. Justify your answer.

If a computer simulation of a map generates a rough orbit, then one wants to know that there is a true orbit of the map which is close to the rough orbit. In this case the true orbit "shadows" the rough orbit. The ability to shadow rough orbits is another kind of stability.

Definition: An ε-chain $y_0, \ldots, y_k$ for a map f is δ-*shadowed* by the orbit of a point x if $d(y_m, f^m(x)) < \delta$ when $0 \leq m \leq k$. A map f has the ε–δ *shadowing property* if any ε-chain can be δ-shadowed. A homeomorphism f is λ-*expansive* if given any two points x and y there exists an integer n such that $d(f^n(x), f^n(y)) > \lambda$.

The following theorem due to Rufus Bowen gives a general method for proving that two homeomorphisms are topologically conjugate.

THEOREM 3.C.1: Suppose that f_1 and f_2 are two homeomorphisms of a locally compact metric space X which satisfy the following conditions:

(a) f_1 and f_2 are 2ρ expansive.
(b) f_1 and f_2 both have the α–ρ shadowing property.
(c) $d(f_1(x), f_2(x)) < \alpha$ for all $x \in X$.

Then f_1 and f_2 are topologically conjugate.

Proof: The f_1-orbit of a point p is an α-chain for f_2. Then there is a point q near p such that the f_2-orbit of q ρ-shadows this chain. Further, q is unique since otherwise f_2 would not be 2ρ expansive. Define a function $h: X \to X$ by $h(p) = q$, where q is the unique point in X whose f_2-orbit ρ-shadows the f_1-orbit of p. Similarly define a function $g: X \to X$ by $g(q) = p$, where p is the unique point whose f_1-orbit ρ-shadows the f_2-orbit of q.

We will show that the functions h and g are inverses of each other. Consider the point $z = g(h(p))$. The f_1-orbit of z ρ-shadows the f_2-orbit of $h(p)$. The f_2-orbit of $h(p)$ ρ-shadows the f_1-orbit of p. Thus the f_1-orbits of both z and p ρ-shadow the same chain and therefore 2ρ-shadow each other. Since f_1 is 2ρ-expansive, $z = p$. By a similar argument, $h(g(q)) = q$. The equation $hf_1 = f_2 h$ follows from the definition of h.

It remains to show that h and g are continuous. By symmetry it is sufficient to show that h is continuous. If h is not continuous, then since X is locally compact and $d(h(p), p) < \rho$, there exists a sequence of points $\{p_n\}$ converging to p such that $\{h(p_n)\}$ converges to $h(w)$ with $w \neq p$. Since f_1 is 2ρ-expansive, there exists an integer m and $\gamma > 0$ such that

$$d(f_1^m(p), f_1^m(w)) > 2\rho + 2\gamma$$

By continuity of f_1 and f_2 there exists a point q of the sequence $\{p_n\}$ such that

$$d(f_1^m(q), f_1^m(p)) < \gamma \quad \text{and} \quad d(f_2^m(h(q)), f_2^m(h(w))) < \gamma$$

From the definition of h we have

$$d(f_1^m(q), f_2^m(g(q))) < \rho \quad \text{and} \quad (f_1^m(w), f_2^m(h(w))) < \rho$$

These four inequalities (see Figure 3.3) imply that

$$d(f_1^m(p), f_1^m(w)) < 2\rho + 2\gamma$$

This is a contradiction and therefore h must be continuous. ∎

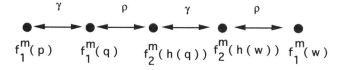

Figure 3.3. Distances between various points.

D. The Hartman–Grobman Theorem

Definition: Two maps f and g are *locally conjugate* if there exist open sets U, V, U^*, V^* and a homeomorphism $h: U \cup V \to U^* \cup V^*$ such that the following diagram commutes:

$$\begin{array}{ccc} U & \xrightarrow{f} & V \\ h \downarrow & & \downarrow h \\ U^* & \xrightarrow{g} & V^* \end{array}$$

Recall that a fixed point p of a C^1 map is *hyperbolic* if the matrix $DF(p)$ is invertible and has no eigenvalues of norm 1.

THEOREM 3.D.1 (The Hartman–Grobman Theorem): If p is a hyperbolic fixed point of a C^1 map $F: R^n \to R^n$, then the map is loally conjugate to its linear part $DF(p)$.

A proof of Theorem 3.D.1 will be given for the case where p is a saddle point for a C^1 map F of the plane. The general result is left as an exercise.

By the linearization procedure, F is topologically conjugate to a map g where

$$g(x) = Ax + E(x) \quad \text{with } E(0) = 0 \text{ and } \|E(x) - E(y)\| \leq \delta \|x - y\|$$

provided that $\|x\| \leq 1$ and $\|y\| \leq 1$. The Lipschitz constant δ can be chosen as small as required. Choose a smooth function

$$r: [0, \infty] \to [0, 1] \quad \text{with } r^{-1}(1) = [0, 0.5] \text{ and } r^{-1}(0) = [1, \infty]$$

Define a new map f by

$$f(x) = Ax + e(x) \quad \text{with } e(x) = r(\|x\|)E(x)$$

Then the map f is locally conjugate to the map F (via an affine map) on the set $\|x\| \leq 0.5$. Thus it is sufficient to prove the following theorem.

THEOREM 3.D.2: There exists $\delta > 0$ such that if

(1) $f(x) = Ax + e(x)$,
(2) A is a diagonal matrix with diagonal elements λ and μ with $0 < |\mu| < 1 < |\lambda|$,
(3) $\|e(x)\| \leq \delta$ and $\|e(x) - e(y)\| \leq \delta \|x - y\|$ for all x and y,

Then f is topologically conjugate to A.

The proof will consist of showing that f and A satisfy the hypothesis of Theorem 3.C.1. Since A is obtained as a special case of f by setting $e(x)$ equal to zero, we must show that f is expansive and has the shadowing property.

LEMMA 3.D.3: For δ sufficiently small, f is a homeomorphism with Lipschitz inverse.

Proof: A^{-1} is a Lipschitz map with constant $k = |\mu|^{-1}$. Hence $\|A(x - y)\| \geq k\|x - y\|$.

$$f(x) - f(y) = Ax - Ay + f(x) - Ax + Ay - f(y)$$

$$\|f(x) - f(y)\| \geq k\|x - y\| - \|e(x) - e(y)\| \geq [k - \delta]\|x - y\|$$

Therefore f has a Lipschitz inverse with Lipschitz constant $[k - \delta]^{-1}$. ∎

PROPOSITION 3.D.4: f is ρ-expansive for any $\rho > 0$.

Proof: Choose δ sufficiently small so that f satisfies the hypothesis of Lemmas 3.B.5, 3.B.6 and Proposition 3.B.7. Also choose δ sufficiently small so that by Lemma 3.D.3 f is a homeomorphism. Suppose that p and q are in a c-horizontal slice. Then $P = f(p)$ and $Q = f(q)$ are a horizontal pair. By Lemma 3.B.5,

$$\|P_1 - Q_1\| \geq k\|p_1 - q_1\| \quad \text{with} \quad k = |\lambda| - \delta(c + 1) > 1$$

Thus, the horizontal distance between $f^n(p)$ and $f^n(q)$ approaches infinity as n approaches infinity.

Now suppose that p and q are not a c-horizontal pair. Then $f^{-n}(p)$ and $f^{-n}(q)$ are also not a c-horizontal pair because f preserves such pairs. If $\|f^{-n}(p) - f^{-n}(q)\| \leq \rho$, then by Lemma 3.B.6, the vertical distance between p and q is less than or equal to $r^n \rho$ with $r = c|\mu| + \delta(1 + c^{-1})$. Since p and q are not a c-horizontal pair, this implies that $\|p - q\| \leq c^{-1} r^n \rho$. Since $r < 1$, there must exist $n > 0$ such that $\|f^{-n}(p) - f^{-n}(q)\| > \rho$. ∎

PROPOSITION 3.D.5: The function f has the $\delta - 1$ shadowing property for δ sufficiently small.

Proof: Let $\{y_j\}$ be an δ-chain for f (see Figure 3.4). Define translations $T_j(x) = x + y_j$ and define $N_j = T_j(N)$. It is sufficient to show that there exists an orbit $\{p_j\}$ of f such that

$$p_j \in N_j \text{ for all integers } j$$

Define functions $f_j = T_{j+1}^{-1} f T_j$. Then from the definitions,

$$f_j(x) = Ax + E_j(x) \quad \text{where} \quad E_j(x) = e(x + y_j) - e(y_j) + f(y_j) - y_{j+1}$$

Furthermore, since e vanishes on the complement of N, the functions E_j all have Lipschitz constant δ and we have

$$\|E_j(0)\| \leq \delta$$

Let $W^s(N, m) = W^s(N, \{f_j\})$, where $j \geq m$. By Theorem 3.B.8, $W^s(N, m)$ is the graph of a Lipschitz function γ_m with Lipschitz constant $1/c$. Since vertical distances contract, by Lemma 3.B.6 there is a unique point z_m in $W^s(N, m)$ which is the intersection of the sets

54 GEOMETRIC METHODS FOR DISCRETE DYNAMICAL SYSTEMS

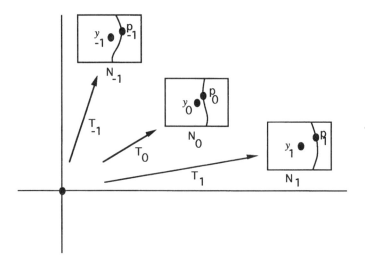

Figure 3.4. Shadowing a δ-chain.

$$W[k, m] = f_{m-1}(\cdots f_k(W^s(N, k))\cdots) \qquad \text{for } k < m$$

The sequence $\{z_j\}$ has the property that $f_j(z_j) = z_{j+1}$ for all j. This sequence is translated into the sequence $\{p_j\}$, where $p_j = z_j + y_j$. From the various definitions,

$$f(p_j) = T_{j+1}(f_j(T_j^{-1}(p_j))) = T_{j+1}(f_j(z_j)) = y_{j+1} + z_{j+1} = p_{j+1}$$

Therefore, as pictured in Figure 3.4, the orbit of p_0 with respect to f shadows the δ-chain $\{y_j\}$. ∎

Proof of Theorem 3.D.2: The functions f and A are both 2ρ-expansive for any ρ by Proposition 3.D.4. They have the $\delta - 1$ shadowing property by Propositions 3.D.5. Thus, by Theorem 3.C.1 they are topologically conjugate. ∎

E. Smale's Horseshoe Map and Symbolic Dynamics

The preceeding three sections of this chapter provide an analysis of the behaviour of orbits near a saddle point. In this section we study an example introduced by Steve Smale to illustrate the behavior of orbits near a complicated invariant set. Invariant sets of this type typically occur in many applications, and Smale's example exhibits some phenomena which are basic to the understanding of chaotic dynamical systems.

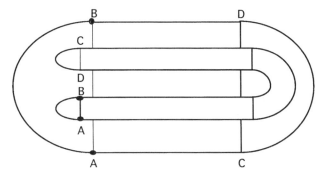

Figure 3.5. The horseshoe map of a stadium-shaped region.

Consider the stadium-shaped region of the plane as pictured in Figure 3.5. Suppose that H is a homeomorphism of the plane that maps the stadium onto the horshoe-shaped region that overlays it. The map H referred to as "horseshoe map."

In Figure 3.6 two copies of the square N with corners A–B–D–C are pictured. The left copy shows the set $H^{-1}(N) \cap N$ as two shaded vertical strips labeled C_0 and C_1. The right copy shows the set $H(N) \cap N$ as two shaded horizontal strips.

To construct our example, suppose that constants a, b, λ, μ are chosen with $0 < \mu < 0.5, 2 < \lambda, 0.5 < a < a + \mu < 1$, and $0.5 < b < b + \lambda^{-1}$. Suppose that the square N and the strips C_0 and C_1 are defined as follows:

$$N = \{x: \|x\| \leq 1\}, \quad C_0 = \{x \in N: |x_1 + b| \leq \lambda^{-1}\}, \quad C_1 = \{x \in N: |x_1 - b| \leq \lambda^{-1}\}$$

Finally, suppose that the restriction of the diffeomorphism H to C_0 and C_1 is given by

$$H(x_1, x_2) = (\lambda[x_1 + b], \mu[x_2 - a]) \quad \text{if} \quad (x_1, x_2) \in C_0$$

$$H(x_1, x_2) = (\lambda[x_1 - b], \mu[x_2 + a]) \quad \text{if} \quad (x_1, x_2) \in C_1$$

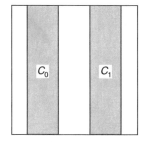 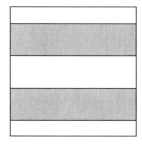

Figure 3.6. Vertical strips map to horizontal strips.

Symbolic Dynamics

Define $\text{Inv}(N) = \{x \in N : H^k(x) \in N \text{ for all integers } k\}$. This set is the largest (maximal) invariant set contained in N. The set $\text{Inv}(N)$ and the behavior of orbits in $\text{Inv}(N)$ can be analysed using what is called "symbolic dynamics." This important technique is introduced here specifically for the horseshoe map. It will be discussed again in a general context in Chapter 4.

Points not in $C_0 \cup C_1$ are mapped outside N. Assign to each point x belonging to $\text{Inv}(N)$ a biinfinite sequence of zeros and ones as follows:

$$s[x,f](j) = \begin{cases} 0 & \text{if } H^j(x) \in C_0 \\ 1 & \text{if } H^j(x) \in C_1 \end{cases}$$

This sequence is called the *itinerary* of the point x.

The *symbol space* Σ is defined to be the set of all biinfinite sequences (or functions)

$$s: Z \to \{0, 1\}$$

The *shift automorphism* $\sigma: \Sigma \to \Sigma$ is defined by the formula

$$\sigma(s)(j) = s(j+1)$$

Define the *itinerary map* $\tau: \text{Inv}(N) \to \Sigma$ by the formula $\tau(x) = s[x,f]$. From these definitions one has the commuting diagram

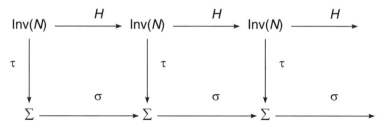

One can see by chasing around the diagram that the function τ takes orbits of H onto orbits of σ.

It is important to make the symbol space into a metric space in a natural way and to establish that the shift automorphism and the itinerary map functions are continuous.

Define a *distance function* on Σ by

$$d(s, t) = \sum_{j=-\infty}^{\infty} 2^{-|j|} |s(j) - t(j)|$$

Exercise: Show that d is a metric on Σ, and show that Σ is a compact metric space with this metric and that it is homeomorhic to the standard

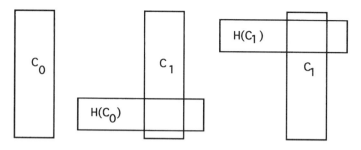

Figure 3.7. Sequence of vertical strips and their images.

Cantor set. Further, show that the shift automorphism σ is a homeomorphism.

The set of all *admissable itineraries* for orbits in Inv(N) is the range of the itinerary map. It is important to characterize this set if possible.

THEOREM 3.E.1: The itinerary map τ associated with the horseshoe H is a homeomorphism.

Proof: Given an itinerary $s \in \Sigma$, draw the sequence of rectangles $\{C_{s(j)}\}$. The map H is affine when restricted to each rectangle, and these rectangles are stretched across each other as pictured in Figure 3.7. For any sequence $\{C_{s(j)}\}$, one can show that there exists exactly one point x in $C_{s(0)}$ whose orbit runs through this sequence of rectangles. The itinerary map takes x onto s. Thus the map τ is onto. It is one-to-one since the point x whose orbit tracks the rectangle sequence is unique. The vertical sides of each rectangle map outside the next rectangle in the sequence. The inverse image of the horizontal sides of each rectangle falls outside the previous rectangle in the sequence. Thus, the orbit of x runs through the interiors of the sequence of rectangles and thus nearby points x and y in Inv(N) track through the same sequence of rectangles for some finite number of iterates. Hence from the definition of the metric on Σ, the distance between the points $\tau(x)$ and $\tau(y)$ is small. This argument shows that the itinerary map is continuous. We leave it to the reader to check that the domain Inv(N) and the range Σ of the itinerary map are compact metric spaces. Consequently, by Proposition A.10 of Appendix A, the inverse of the itinerary map is continuous and τ is a homeomorphism. Further, from the definition of τ, we have $\tau H = H\tau$ and so the map H restricted to Inv(N) is topologically conjugate to the shift automorphism.

COROLLARY 3.E.2: Periodic points are dense in Inv(N) and there are exactly 2^m periodic points of period less than or equal to m in Inv(N).

Proof: There are exactly 2^m periodic points of period less than or equal to m in Σ, since there are this many distinct sequences of zeros and ones of length m. Further, any sequence in Σ may be approximated by choosing a periodic sequence which matches the given sequence exactly for $-k \leq j \leq k$. Using the definition of the metric on Σ, one gets an estimate of the distance between the approximating periodic sequence and the given one. Thus, periodic points are dense in Σ.

Since H is topologically conjugate to σ, there is a one-to-one correspondence between periodic points in Inv(N) and periodic points in Σ. Therefore, periodic points are dense in Inv(N) and there are exactly 2^m periodic points of period less than or equal to m in Inv(N). ∎

F. Hyperbolic Invariant Sets

Terminology developed to study differential topology is necessary to formulate the results of this section. For a brief discussion of tangent bundles, tangent maps, and diffeomorphisms, see Appendix C. The invariant set Inv(N) for the horseshoe map is an example of a hyperbolic invariant set. These types of invariant sets have very strong stability properties.

Definition: Suppose that $F: R^n \to R^n$ is a C^r map, and suppose that Λ is a compact invariant set for f. The set Λ is a *hyperbolic invariant set* if there exists $b > 0$ and integer $m > 0$, and there exists an isomorphism $\phi_x: V^n \to x \times V^n$ for each x in Λ, such that

$$\|\phi_x\| \leq b, \quad \|\phi_x^{-1}\| \leq b.$$

Define $A_x: V^n \to V^n$; $A_x = \phi_{f(x)}^{-1} Df^m(x) \phi_x$. The linear transformation A_x must satisfy the following conditions: There exits constants $0 < \mu < 1 < \lambda$, and there exists norms $\|\ \|_s$ on V^s and $\|\ \|_u$ on V^u where $V^u \times V^s = V^n$ such that if $v = (v_1, v_2) \in V^u \times V^s$ then

(1) $\|v_2\|_s \leq 2\|v_1\|_u$ implies that $\|A_x v\|_u \geq \lambda \|v\|_u$;
(2) $\|v_2\|_s \leq 2\|v_1\|_u$ implies that $\|w_2\|_s \leq 2\|w_1\|_u$ where $A_x v = (w_1, w_2)$;
(3) $\|v_2\|_s \geq 2\|v_1\|_u$ implies that $\|w_2\|_s \leq \mu \|v_2\|_u$ where $A_x v = (w_1, w_2)$.

Example: Consider the horseshoe map defined in section 3.E. From the definition of H we have

$$DH(x) = A = \begin{pmatrix} \lambda & 0 \\ 0 & \mu \end{pmatrix}$$

at each point $x \in C_0 \cup C_1$. To show that Inv(N) is a hyperbolic invariant set, for each point x in Inv(N) define the isomorphism

$$\phi_x: V^2 \to x \times V^2; \qquad \phi_x(v) = (x, v)$$

From the definition, the associated isomorphism A_x is equal to A. The box norm is used on V^2 and the other norms are defined to be the absolute values on the components of a vector in V^2. The three conditions for the definition of hyperbolicity are satisfied.

The following theorem states a selection of advances results that hold for hyperbolic invariant sets. Proofs of these and related results can be found in *Global Stability of Dynamical Systems* by Mike Shub (1987). See also Conley (1975). A proof can probably be given along the lines of our proof of the Hartman–Grobman theorem. The idea is to cover the invariant set by boxes which are mapped across each other in a nearly affine manner.

THEOREM 3.F.1: Let Λ be a hyperbolic invariant set for a diffeomorphism f of R^n. Then

(1) The restriction of f is Λ is expansive.
(2) Given $\rho > 0$ there exists $\alpha > 0$ such that any ρ-chain can be α shadowed.
(3) There exists a compact neighborhood U of Λ such that any invariant set contained in U is hyperbolic.
(4) There exists a compact neighborhood U of Λ and a neighborhood V of f in the C^1 topology on diffeomorphisms from U into R^n such that for any g in V there exists a hyperbolic invariant set Λ_g for g and a topological conjugacy $h: \Lambda \to \Lambda_g$ taking f-orbits onto g-orbits.
(5) If x is a chain recurrent point in Λ then x is the limit of hyperbolic periodic points.
(6) If x is a chain recurrent point in Λ which is not periodic, then given any neighborhood of x there exists an integer m and an invariant set J of f^m and a topological conjugacy between the restriction of f^m to J and the shift automorphism on Σ.

G. Trellis Structure and Resonance Zones

The set of all points whose orbits converge to a fixed point p is called the *stable manifold* of p. For a smooth map of the plane we have shown (Theorem 3.B.3) that the stable manifold of a saddle point contains a smooth curve passing through the point. In this section we study the global behavior of the stable and unstable manifolds of a saddle point p of a diffeomorphism f of the plane. The stable manifold of the saddle point is a smooth curve in the plane which passes through p and does not cross itself. Similarly, the unstable manifold of p is also such a smooth curve. However, the stable and unstable manifolds may intersect.

In chapter 33 of *New Methods in Celestial Mechanics*, Poincaré (1899) describes the figure formed by the stable and unstable manifolds of a saddle point as follows:

> When we try to represent the figure formed by these two curves and their intersections, each of which corresponds to a doubly asymptotic solution, these intersections form a type of trellis, tissue, or grid with infinitely serrated mesh. Neither of these two curves must ever cut across itself again, but must bend back upon itself in a very complex manner in order to cut across all of the meshes in the grid an infinite number of times. The complexity of this figure will be striking, and I shall not even try to draw it. Nothing is more suitable for providing us with an idea of the complex nature of the three-body problem, and all of the problems of dynamics in general.

Following Poincaré, the figure formed by the stable and unstable manifolds of a finite collection of hyperbolic periodic points will be called a *trellis*. This figure is also commonly referred to as a *homoclinic tangle*. Figure 1.6 shows the beginning development of the trellis of a saddle point $p = (-1.9, -1.9)$ for the Hénon map $H(x, y) = (y, -0.9x - y^2)$. In this section we will analyze and describe certain trellis structures.

Definition: Let f be a diffeomorphism of the plane having a saddle point p. The *stable and unstable manifolds* of p are

$$W^s(p,f) = \{x \in R^2 : f^n(x) \to p \text{ as } n \to \infty\}$$
$$W^u(p,f) = \{x \in R^2 : f^n(x) \to p \text{ as } \to -\infty\}$$

By the linearization procedure, there is an affine map L of the plane which maps the origin to p and maps the unit square N (in the box norm) onto a compact neighborhood B of the fixed point p. Further, by Theorem 3.B.3, there is a smooth function $\lambda^s : [-1, 1] \to [-1, 1]$ such that $W^s(B,f) = \{L(\lambda^s(t), t) : -1 \le t \le 1\}$. Define a smooth curve $\gamma^s : [-1, 1] \to R^2$ by $\gamma^s(t) = L(\lambda^s(t), t)$. Note that the tengent vector $\dot\lambda(t) \ne 0$. A function with this property is called an *immersion*. The range of this function is the set $W^s(B,f)$, which we refer to as the *local stable manifold* of p.

PROPOSITION 3.G.1: The stable manifold of the saddle point p is the union of all preimages of the local stable manifold of p. Thus, $W^s(p,f) = \cup \{f^n(W^s(B,f)) : n \le 0\}$. Furthermore, there is an extension of the immersion γ^s to an injective immersion of the real line whose range is $W^s(p,f)$. Thus, the stable manifold is a smooth curve which does not cross itself.

Proof: If $x \in W^s(p,f)$ then, since the iterates of x converge to p and B contains p in its interior, there is an integer m such that $f^m(x) \in W^s(B,f)$. Thus, $x \in f^{-m}(W^s(B,f))$. The dynamics of f on $W^s(B,f)$ induces dynamics on $[-1, 1]$. The induced map ϕ is defined by $\phi(t) = (\gamma^s)^{-1}f\gamma^s(t)$. To extend the immersion γ^s, we first choose some appropriate constant $k < 1$ and extend ϕ to a smooth monotone function on the real line with $|\phi(t)| \leq k|t|$ when $|t| \geq 1$. This ensures that all orbits of ϕ converge to zero. For a real number t let $m(t)$ be the least integer such that $\phi^{m(t)}(t) \in [-1, 1]$. Define $\gamma^s(t) = f^{-m(t)}(\gamma^s(\phi^{m(t)}(t)))$. To show that the extension is one-to-one, suppose that $x = \gamma^s(t) = \gamma^s(r)$. Then $m(t) = m(r)$ because $m(t)$ is exactly the least integer with $f^{m(t)}(x) \in B$. Since the functions $f^{-m(t)}$, γ^s, $\phi^{m(t)}$ used to define $\gamma^s(t)$ and $\gamma^s(r)$ are one-to-one we must have $t = r$. We leave it to the reader to check that the extended function γ^s is an immersion. ∎

Definition: Suppose that g is a map of a metric space X and assume that z is a fixed point of this map. A point $q \neq z$ in X is a *homoclinic to z* if $g^n(q) \to z$ as $n \to \infty$ and there exists a preorbit $\{q_n\}$ for q such that $q_n \to z$ as $n \to -\infty$. If p is a saddle point for a diffeomorphism f of the plane, then a point $q \in W^s(p,f) \cap W^u(p,f)$ is called a *transverse homoclinic point* if the tangent vectors to the smooth curves $W^s(p,f)$ and $W^u(p,f)$ at p are linearly independent.

Remark: From now on, we will assume that the Jacobian matrix $Df(p)$ at the saddle point has real eigenvalues $0 < \mu < 1 < \lambda$. This does not result in a loss of generality since $W^s(p,f) = W^s(p,f^2)$ and one may replace f by f^2 if necessary. The reason for the assumption to ensure that the map f does not interchange the two components of the set $W^s(p,f) - \{p\}$, and also does not interchange the two components of the set $W^u(p,f) - \{p\}$.

Terminology: Choose immersions γ^s, $\gamma^u: R^1 \to R^2$ such that $\gamma^s(0) = \gamma^u(0) = p$ whose images are the stable and unstable manifolds of p, respectively. The linear ordering on the real numbers produces corresponding linear orderings $<_s$ and $<_u$ on these manifolds called the *stable and the unstable orderings*. Given points a and b in W^s, the closed segment of stable manifold between these points will be denoted by $S[a, b]$. Given points c and d in W^u, the closed segment of unstable manifold between these points will be denoted by $U[c, d]$. Let $H = W^s \cap W^u$ denote the set of points homoclinic to the saddle point p. The set of homoclinic points H is linearly ordered by both the stable and the unstable orderings. The set H together with these orderings contains a lot of information about the trellis associated with the saddle point.

Definition: A *primary homoclinic point* is a homoclinic point q such that $S[p, q] \cap U[p, q] = \{p, q\}$. A *fundamental segment* is a segment of stable or unstable manifold of the form $S[f(q), q]$ or $U[q, f(q)]$ where q is a primary homoclinic point. An *initial segment* of stable or unstable manifold is a segment of the form $S[p, a]$ or $U[p, c]$. A homoclinic point q is of *type k* if

$$S[p, f^k(q)] \cap U[p, q] - \{p\} = \emptyset$$

and

$$S[pf^j(q)] \cap U[p, q] - \{p\} \neq \emptyset \quad \text{when} \quad 0 < j < k$$

The type number of a homoclinic point is also sometimes called its Birkhoff signature. Note that the orbit of a homoclinic point q of type k consists of homoclinic points having the same type number. A primary homoclinic point has type number equal to 1. Trellises also have different geometric types, and it is an interesting open problem to classify the different types of trellises that may occur (Easton, 1985; Rom-Kedar, 1994). Equivalent trellises should at least have sets of homoclinic points which are homeomorphic via a homeomorphism that preserves both the stable and unstable orderings. For example, if one trellis has homoclinic points of type m and another trellis does not, then they are distinct.

Definition: A *resonance zone* is a region bounded by alternating initial segments of unstable and stable manifolds of saddle points which intersect only in their endpoints.

Resonance zones are used to localize the study of trellises. One wants to understand the behavior of orbits inside a resonance zone. Two examples will be analyzed below.

Definition: The parts of the stable and unstable manifolds of p which are localized in a resonance zone R are called the R-*stable and R-unstable manifolds of p*. They are defined to be the sets

$$W^s(p, R) = \{x \in W^s(p) : f^n(p) \in R \text{ for all } n \geq 0\}$$
$$W^u(p, R) = \{x \in W^u(p) : f^n(p) \in R \text{ for all } n \leq 0\}$$

Example: Figure 3.8 pictures the beginning development of a trellis associated with the horseshoe map. There is a saddle point p located in the intersection of the strip C_0 (pictured in Figure 3.5) and its image. The unstable manifold of p starts out as a horizontal line segment which crosses the square in Figure 3.5. The stable manifold of p starts out as a vertical line segment which crosses from the bottom to the top of the strip C_0. A resonance zone R pictured in Figure 3.8 is defined to be the region bounded by the vertical initial segment $S[p, a_1]$ and the horseshoe-shaped initial segment $U[p, a_1]$. The subscripts are used to indicate that $H(a_j) = a_{j+1}$.

The exit set E for R is the D-shaped region bounded by the segments $S[a_0, b_0]$ and $U[a_0, b_0]$ and the entry set for R is the D-shaped region bounded by the segments $S[b_0, a_1]$ and $U[b_0, a_1]$. A critical observation about the horseshoe map is that the image of $R - E$ is the horseshoe-shaped

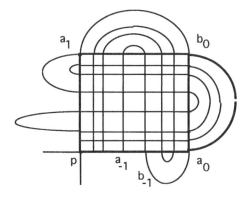

Figure 3.8. Trellis associated with the horseshoe map.

region which is pictured in Figure 3.9. The R-unstable manifold can be recursively constructed by mapping $R - E$ into R.

The unstable segment $U[b_0, a_1]$ maps to the segment $U[b_1 a_2]$ which intersects the square $R - E$ in two horizontal line segments. The first two stages of the construction are pictured in Figure 3.8. As this process continues, at the nth stage one can see that the R-unstable manifold crosses this square in a family of 2^n horizontal line segments whose right endpoints form the endpoints of removed open invervals for a Cantor set under construction on the segment $S[b_0, a_0]$. These endpoints are connected by semicircular segments of the unstable manifold which are contained in the exit set. Thus the set $W^u(p, R)$ consists of a countable union of nested D-shaped segments roughly parallel to each other. The innermost segment is $U[b_1, a_2]$ and the outermost segment is $U[p, a]$. The closure of the set $W^u(p, R)$ intersects the vertical segment $S[b_0, a_0]$ in a Cantor set.

A second type of trellis is formed when the segment $U[b_1, a_2]$ does not intersect the segment $S[b_0, a_0]$, but its image $U[b_2, a_3]$ does intersect the segment $S[b_0, a_0]$, as illustrated in Figure 3.10. Note that in this situation it takes exactly two iterates before the image of the entrance set intersects the exit set of the resonance zone. A third type of trellis will be formed when this type of transfer from entrance to exit set takes three iterates. Thus an infinite number of trellises of different types as possible, each with their own combinatorial structures. The horseshoe trellis is a trellis of type 1. Next we will construct a model of a trellis of type 2.

Example: The Hénon map $H(x, y) = (2 - y - x^2, x)$ is used to generate the trellis of type 2 pictured in Figure 3.11. The point Q is a fixed point, but is not a saddle point. Rather, it is an "elliptic" fixed point and the linear approximation to the map H near Q is a rigid counterclockwise rotation. The right side of the unstable manifold of p begins with the segment

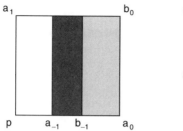 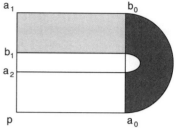

Figure 3.9. The left square is the set $R - E$. The image of the three vertical strips overlaps the set $R - E$ is pictured on the right.

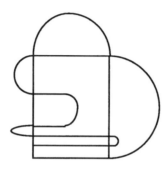

Figure 3.10. A trellis of type 2.

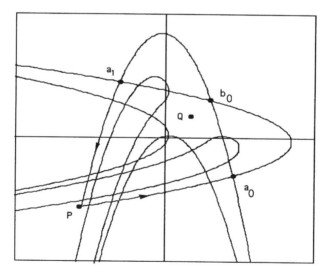

Figure 3.11. Trellis for the Hénon map $H(x, y) = (2 - y - x^2, x)$. The heart-shaped region bounded by segments of the stable and unstable manifolds of P is a resonance zone.

$U[p, a_0]$, The picture was drawn using a computer to approximate the stable and unstable manifolds of p. The window pictured is a square of side 10 centered at the origin. Only the top half of the stable manifold rising above p was drawn. Similarly, only the right half of the unstable manifold was plotted, and only the intersection of these manifolds with the square is pictured. The map H is area-preserving since the determinant of its Jacobian matrix at any point in the plane is equal to 1. Area-preserving maps will be discussed in Chapters 6 and 7.

The resonance zone that we will use is heart-shaped region R pictured in Figure 3.11 which is bounded by the initial segments $U[p, b_0]$ and $S[p, b_0]$. The exit set for this zone is bounded by the segments $U[a_0, b_0]$ and $S[b_0, a_0]$. One may imagine coloring points in the resonance zone according to their forward or backward exit times. In this case, the color would change only upon crossing a stable or unstable manifold segment. The reason is that discontinuities of the exit time functions occur only at points whose orbits intersect the boundary of the resonance zone. Since the boundary is composed of segments of stable and unstable manifolds, the discontinuity points of the forward and backward exit time functions must also belong to the R-stable and R-unstable manifolds, respectively.

Our goal is to describe the R-unstable manifold of p. It is formed as follows: Delete the part of the initial segment $U[p, b_0]$ that intersects the exit set. Apply the map to the resulting segment $U[p, a_0]$, then delete the part of the image that intersects the exit set. The connected components of the clipped image are called *strings*. Apply the map to each string, then form new strings by deleting the intersection of their images with the exit set. Continue to stretch and clip each string. Thus the local unstable manifold internal to the resonance zone is produced as the union of images of strings. Similarly, the local stable manifold of R is generated by using the inverse of the map to clip and stretch the segment $S[p, b_0]$. However, for the map H there is a short cut. The reflection map of the plane $S(x, y) = (y, x)$ transforms the unstable manifold of p to the stable manifold of p, because (as one may verify) we have $H^{-1} = SHS$.

It is instructive to carry out the first few steps of the construction of the R-unstable manifold. The image of the string $U[b_0, a_1]$ is the string $U[b_1, a_2]$. Note that the lobe bounded by the segments $U[b_1, a_2]$ and $S[a_2, b_1]$ is the image of the entry set of R. Thus, the unstable manifold $W^u[p, R]$ does not intersect the interior of this set, and therefore all strings are contained in the region Λ pictured in Figure 3.12, which is bounded by the following alternating segments of unstable and stable manifolds:

$$U[p, a_0], \quad S[a_0, b_0], \quad U[b_0, a_1], \quad S[a_1, b_1], \quad U[b_1, a_2], \quad S[a_2, p]$$

The endpoints of each string must belong to the three segments of stable manifold on the boundary of Λ. There are three types of strings. Those that join $S[p, a_2]$ to $S[b_0, a_0]$ are called *alpha strings*. Those that join $S[b_0, a_0]$ to

$S[b_1, a_1]$ are called *beta strings*, and those that join $S[b_1, a_1]$ to $S[p, a_2]$ are called *gamma strings*.

Each string produces a shower of fragments as it is stretched by the map H and clipped into pieces by deleting its intersection with the exit set. One may think of strings as forming a population. Each alpha string produces as "offspring" one alpha and one beta string. Each beta string produces one gamma string, and each gamma string produces two alpha strings.

$$\alpha \to \alpha + \beta, \qquad \beta \to \gamma, \qquad \gamma \to \alpha + \alpha$$

A *string population model* can be formed by defining a population vector $v(n)$ with three components specifying the number of alpha, beta and gamma strings present at the nth generation. Then $v(n+1) = Av(n)$, where A is the matrix

$$\begin{pmatrix} 1 & 0 & 2 \\ 1 & 0 & 0 \\ 0 & 1 & 0 \end{pmatrix}$$

The population of fragments of the (primordial) gamma string is given by $A^n v(0)$, where $v(0)$ is the column vector with entries $(0, 0, 1)$. This collection constitutes the R-unstable manifold of p which we seek to describe.

PROPOSITION 3.G.2: The asymptotic growth rate of the string population is equal to the largest eigenvalue λ of the matrix A. The value of λ is approximately 1.6956.

Proof: The number of strings at the nth stage in the construction is the sum of the entries in the vector $A^n v(0)$. The matrix A produces a linear map whose dynamics is determined by the eigenvalues and eigenvetors of the matrix. One can write $v(0)$ as a linear combination of the eigenvectors of A. The growth rate of strings is $\lim_{n \to \infty} \ln(\|A^n v(0)\|)/n$, and this limit is clearly equal to the largest eigenvalue of A. ∎

The left endpoints of the alpha strings have a least upper bound e on the stable manifold of P. Similarly, the lower endpoints of the gamma strings have a greatest lower bound E on the stable manifold of P. Since distinct strings can never cross, we have $e \leq E$ in the ordering along the stable manifold. Suppose that $e = E$. Then there should be a separatrix between the alpha and gamma strings. Similarly, there may be a separatrix between alpha and beta strings, and between beta and gamma strings as pictured in Figure 3.12.

Since beta strings map to gamma strings and gamma strings map to alpha strings, therefore the point F should map to G and the point G should map to E. This scenario suggests that the separatrices pictured might be formed by the unstable manifold of a period-3 saddle. Solving the equation $H^3(x, x) = (x, x)$ is equivalent to solving the equation $x^4 + 2x^3 - 3x^2 - 2x + 2 = 0$, which has

HYPERBOLIC FIXED POINTS

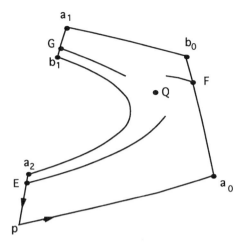

Figure 3.12. Separatrices separating different types of strings.

$x = 1$ as a solution. Hence, the period-3 point Z has coordinates $(1, 1)$. A computer drawing of the stable and unstable manifolds of Z and its iterates is pictured in Figure 3.13. The stable and unstable manifolds of Z are only partially drawn and they continue to weave through the picture in a complex pattern which we seek to describe.

The period-3 point Z and its stable and unstable manifolds is key to forming a model for a trellis of type 2. Temporarily we assume that the triangular region around the elliptic fixed point Q is formed by the nontransverse intersection of stable and unstable manifolds of this orbit. This assumption is false even though this is not evident at the level of resolution shown in Figure 3.13. Now consider map F, which agrees with the map H away from a small neighborhood of the triangle, and suppose that the map F^3 leaves the triangle invariant. One can calculate where various segments of stable and unstable manifold transform relative to the map F, and the resulting picture (hand-drawn this time) is Figure 3.14. One recognizes a horseshoe map on three symbols is the result of mapping the region X with corners p, a_0, F, Z, Z_2, E across itself with F^3. The image of X is the shaded region.

The region X is topologically a rectangle, and the shaded image $F^3(X)$ crosses X in three strips. Thus we have redrawn the figure as shown in Figure 3.15.

The boundaries of the horizontal strips are formed by the unstable manifolds of P, and Z and the stable manifold of Z_2. To distinguish these manifolds, suppose they are colored red, yellow, and orange respectively. The pattern of their intersections with the segment of stable manifold $S[p, E]$ is represented by the work ROYRRO. This is the first stage in the construction of a Cantor set based on removing two open intervals at each stage of the construction. At the second stage of the construction the pattern repeats in the first and third intervals and reverses in the second interval. Thus, the word representing the ordering of the homoclinic and heteroclinic intersection on the segment $S[p, E]$ is ROYRRO–YRRYOR–ROYRRO. The combinatorial structure of the R-unstable manifolds (with respect to F) of P, Z_2 and Z, respectively, within the region X is nonintersecting horizontal line segments whose intersections with the segment of stable manifold $S[p, E]$ form the endpoints of a Cantor set constructed as above.

By transforming the region X by F one extends the information about the structure of the set $W^u(p, R) \cap X$ to more of R. Eventually the full picture emerges. However, rather than pursuing this rather complicated construction we conclude the discussion with a few remarks about the utility of resonance zones.

The usual way to think about resonance is by analogy with the build-up in the amplitude of oscillations in a periodically forced oscillator such as a pendulum. More abstractly, one may define resonance in discrete dynamical systems as the escape from a neighborhood of a fixed point. This is of special interest near a fixed point such as Q where the linear approximation to the map indicates that nearby points may not escape from a neighborhood of the point. Then the nonlinear terms in the map determine the outcome. For the Hénon map H in the example above one can show that all orbits which start outside

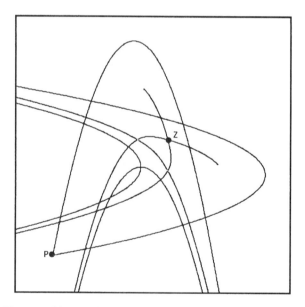

Figure 3.13. The unstable manifolds of a period-3 orbit form the separatrices between strings. The point Z is on the periodic orbit. The stable and unstable manifolds of Z are only partly drawn. Note that the unstable manifolds of P and Z cannot cross each other. Likewise, the stable manifolds of these points cannot cross.

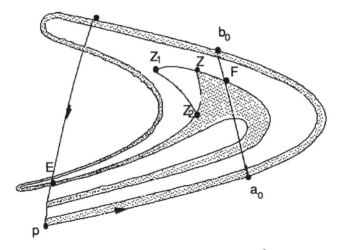

Figure 3.14. The shaded region is the set $F^3(X)$.

the resonance zone are unbounded either forward of backward. Thus the most interesting orbit structure is contained in the resonance zone.

The partitioning of a resonance zone R into pieces bounded by alternating segments of stable and unstable manifolds gives information about how these pieces are transported through the zone since their boundaries transform in a predictable manner. When the map perserves area, questions regarding transport of areas through a resonance zone have applications to problems arising in fluid dynamics and chemistry (Wiggins, 1992).

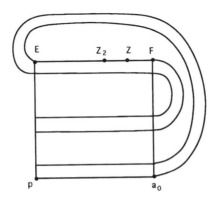

Figure 3.15. Redrawn figure of $F^3(X)$ crossing X.

H. Topological Entropy

One measure of the complexity of a map is the growth rate of the number of "different" orbit segments of length n as n increases. This is formalized by the concept of topological entropy discussed below. Using the ideas of resonance zones and strings discussed in the preceeding section, one can sometimes estimate the topological entropy of a map.

Definition: Let $f: X \to X$ be a map of compact metric space X. A subset E of X is n–ε *separated* if for any pair of distinct points x and y both belonging to E there exists an integer j with $0 \leq j < n$ such that $d(f^j(x), f^j(y)) > \varepsilon$. Let $Sn(n, \varepsilon)$ be the number of points in the largest n–ε-separated set and define

$$h(f, \varepsilon) = \limsup_{n \to \infty} \ln(S(n, \varepsilon)/n) \quad \text{and} \quad h(f) = \lim_{\varepsilon \to 0} h(f, \varepsilon)$$

The number $h(f)$ is called the *topological entropy* of the map f.

Exercise: Shown that topologically equivalent maps must have the same topological entropy.

Definition: A subset E of X is n–ε-*chain separated* if for any two distinct points x and y belonging to E there exists an integer j with $0 \leq j < n$ and ε-chains starting at x and y, respectively, such that $d(x_j, y_j) > \varepsilon$. Let $S(n, \varepsilon, \varepsilon)$ be the number of points in the largest n–ε-chain-separated set and define the *chain entropy* of the map to be the number

$$h(f, \varepsilon, \varepsilon) = \limsup_{n \to \infty} \ln(S(n, \varepsilon, \varepsilon)/n)$$

THEOREM 3.H.1: The topological entropy of a map is the limit of the chain entropies:

$$h(f) = \lim_{\varepsilon \to 0} h(f, \varepsilon, \varepsilon)$$

THEOREM 3.H.2: If the map f is expansive, and has the shadowing property, then $h(f)$ is the growth rate of the number of periodic points:

$$h(f) = \lim_{k \to \infty} N(f, k)/k$$

where $N(f, k)$ is the number of periodic points of f having period less than or equal to k.

Theorems 3.H.1 and 3.H.2 are proved in Barge and Swanson (1990).

Application: For the horseshoe map H, we have calculated that $N(H, k) = 2^k$. Therefore, by the Theorem 3.H.2, the topological entropy of H is $\ln(2)$.

Example: This is an estimate for the topological entropy of the Hénon map discussed at the end of section G. Start with an α-string. The asymptotic growth rate of the number of string fragments is the largest eigenvalue of the matrix A. To see that this is a lower bound for the topological entropy, assign an itinerary to each point in the original string. The word assigned to a point x is the sequence of types of strings that it belongs to at the jth iterate for $0 \leq j \leq k$. The endpoints of α-, β-, γ-strings are ε-separated and therefore there is a k–ε-separated set of size approximately λ^k contained in the original α-string. Thus, a lower bound for the topological entropy is $\ln(\lambda)$.

I. Problems

1. Show that the matrix norm assiciated with the box norm for a 2×2 matrix M is $||M|| = \max\{|M_{11}| + |M_{12}|, |M_{21}| + |M_{22}|\}$.
2. If M is an $n \times n$ matrix with all eignevalues of norm less than 1, define a norm on R^n such that the associated matrix norm of M is less than 1.
3. Generalize and prove Theorem 3.B.1 (the topological stable manifold theorem) in d dimensions. To formulate the theorem, modify the hypothesis of Theorem 3.B.15. You will need to use the result from topology that there is no retraction of a solid n-dimensional ball onto its boundary (See the hint for Problem 4.)
4. Generalize the proof of Theorem 3.B.2 to d dimensions. Hint: In generalizing Proposition 3.B.7 you will need to show that the image of a horizontal slice intersects N in a new horizontal slice. The projection of the new slice must be equal to the projection of N. The argument requires the use of an important theorem from topology: It is impossible to have retraction of the k-dimensional disk onto its boundary. The disk is the set of points in R^k whose distance from the origin is less than or equal to 1. Its boundary is the $(k-1)$-dimensional sphere consisting of all points exactly distance 1 from the origin. A *retraction* from a space X onto a subspace A is a continuous function f from X onto A such that $r(x) = x$ for all points x belonging to A.
5. Prove Theorem 3.B.15.
6. Generalize the proof of the Hartman–Grobman theorem to d dimensions.
7. Find a logistic map which is not structurally stable within the family of logistic maps. Justify your answer.
8. Find a logistic map which is structurally stable within the family of logistic maps. Justify your answer.
9. Complete the proof of Theorem 3.E.1 by showing that the itinerary map is continuous. Hint: Show that orbits of points in the invariant set track through the interiors of the vertical strips.

10. Compute the topological entropy of logistic maps for selected parameter values.
11. A map f is *semi-conjugate* to a map g if there exists a continuous function ϕ such that $\phi f = g\phi$. If this is so, show that the topological entropy of f is greater than or equal to the topological entropy of g.
12. Find and present proofs of Theorem 3.H.1 and 3.H.2.

J. Further Reading

There are several good treatments of the stable manifold theorems which are different from the one presented in this chapter. The reader may consult Shub (1987), Robinson (1995), and McGehee (1973). For more information on trellis structure and resonance zones, see Easton (1985), Easton (1991), Easton (1993), Rom-Kedar and Wiggins (1988), and Rom-Kedar (1994).

4

Isolated Invariant Sets and Isolating Blocks

For a map f of a metric space X, one can ask the basic questions: How does one locate the invariant sets of f? When does a particular compact subset Y of X contain an invariant set? In this chapter, we partially answer these questions by using compact sets with special properties called isolating blocks to locate and study special invariant sets called isolated invariant sets.

Definition: An *isolating region* is a compact set R such that every invariant set contained in R is contained in the interior of R. An *isolated invariant set* is the maximal invariant set contained in some isolating region. An *isolating block* is a compact set N such that whenever three points $x, f(x), f^2(x)$ on an orbit are contained in N, then the middle point $f(x)$ is contained in the interior of N.

It is easy to see that an isolating block is an isolating region and that not all isolating regions are isolating blocks. We will show that each isolated invariant set is the maximal invariant set contained inside an associated isolating block. One can use the topology of a block and the way it maps across itself to prove that it isolates a nonempty invariant set. Further properties of this isolated invariant set can sometimes be deduced using symbolic dynamics. We give a general construction of isolating blocks using epsilon chains. Isolating blocks are stable with respect to perturbations of the map and therefore properties of the dynamics which can be deduced from blocks persist.

A. Attracting Sets

Roughly, an attracting set is an invariant set with the property that all orbits which start in some neighborhood of the set converge to it. Attracting sets have several types of stability properties. A property called "chain-stability" ensures that the attracting set is observable in a computer simulation of the dynamics provided that round-off errors in the computation are sufficiently small.

Definition: A compact set N is an *attractor block* (or *trapping region*) for f if $f(N)$ is contained in the interior of N. An attractor block is a special type of isolating block. A compact invariant set A is an *attracting set* if A is the maximal invariant set contained in some attractor block N. The stable set $W^s(A)$ of an attracting set A is called its *basin of attraction*.

If N is an attractor block, then by Theorem 2.B.5, $\omega(N,f)$ is the maximal invariant set contained in N. Thus each attracting set A has an attractor block N such that $A = \cap \{f^n(N) : n \geq 0\}$.

Example: The *solenoid* (see Figure 4.1) is an attracting set for a map of R^3 which transforms a solid torus inside itself in so that the image of the center-line wraps twice around the torus. The image of the nth iterate of the torus intersects a cross section of the toris in a set of 2^n disjoint disks. From this one can show that the solenoid intersects a cross section in a Cantor set. Thus the solenoid is a fairly complicated set, a "strange" attractor.

Definition A compact invariant set S is

1. *Stable* if given a neighborhood U of S there is a neighborhood V of S such that $f^n(V)$ is contained in U for all $n \geq 0$.
2. *Limit stable* if S is contained in the interior of $W^s(S)$.
3. *Asymptotically stable* if it is both stable and limit stable.
4. *Chain stable* if given a neighborhood U of S there exists an $\varepsilon > 0$ such that $\text{ch}(S, X, \varepsilon) \subset U$.

Example: Let g be a smooth map of the unit circle which has exactly one fixed point at $(1, 0)$. Suppose that g moves all the other points counterclockwise around the circle. Then the fixed point is limit stable, but is not stable. For the identity map of the unit circle, all fixed points are stable, but not limit stable.

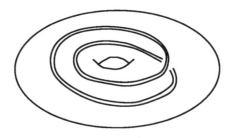

Figure 4.1. The solenoid mapping of a solid torus twice around itself. The intersection of successive images is a strange attractor.

ISOLATED INVARIANT SETS AND ISOLATING BLOCKS 75

The following elegant construction of attractor blocks is due to R. McGehee (1988).

PROPOSITION 4.A.1: If S is a subset of X and $N = \text{ch}(S, X, \varepsilon)$ is compact and $X - N \neq \varnothing$, then N is an attractor block. Further, $d(f(N), X - N) \geq \varepsilon$.

Proof: Suppose that $d(f(N), X - N) < \varepsilon$. Then there exist points y belonging to N and z belonging to $X - N$ with $d(f(y), z) < \varepsilon$. Choose a neighborhood W of y such that $d(f(w), z) < \varepsilon$ for each point w in W. Since y belongs to N, there exists an epsilon-chain $c_0, \ldots, c_k$ with c_0 in S and c_k in W. Thus $c_0, \ldots, c_k, z$ is an epsilon-chain, and this implies that z belongs to N. This is a contradiction. Therefore, $d(f(N), X - N) \geq \varepsilon$, and N is an attractor block. ∎

PROPOSITION 4.A.2: If S is an attracting set, then S is chain stable.

Proof: Choose an attractor block N of S. Given a neighborhood U of S, by Theorem 2.B.5 there exists an integer m such that $f^m(N) \subset U$. The set $B = f^m(N)$ is also an attractor block for S. Choose $\varepsilon > 0$ such that $d(f(B), X - B) > \varepsilon$. Then by the choice of ε, $\text{ch}(S, X, \varepsilon)$ is contained in B and hence in U. Therefore, S is chain stable. ∎

PROPOSITION 4.A.3: If S is asymptotically stable, then there exists a compact neighborhood V of S contained in $W^s(S)$ and a positive integer m such that V is an attractor block for f^m.

Proof: Since S is asymptotically stable, there exists a compact neighborhood V of S which is contained in $W^s(S)$. Since S is stable, there exists a compact neighborhood W of S such that $f^j(W)$ is contained in the interior of V for all j. Since V is contained in the stable set of S, for each point x of V one can find an integer $m(x)$ and a neighborhood U_x of x such that $f^{m(x)}(U_x) \subset W$. The set V is covered by finitely many such neighborhoods. Let m be the maximum of the finitely many corresponding $m(x)$. Since $f^j(W)$ is contained in the interior of V for all $j \geq 0$, we have $f^m(V)$ contained in the interior of V. Hence, V is an attractor block for f^m. Since V is a compact subset of $W^s(S)$, it follows that the intersection of the sets $f^j(V)$ is equal to S. ∎

PROPOSITION 4.A.4: Given a compact set K, a positive integer m, and $\varepsilon > 0$, there exists $\delta > 0$ such that if $y_0, \ldots, y_m$ is a δ-chain in K, and $f^j(y_0) \in K$ for $0 \leq j \leq m$, then $d(f^m(y_0), y_m) < \varepsilon$.

Proof: The proposition is true for $m = 1$. Assume it is true for m. Choose $0 < \delta < \varepsilon/2$ such that $d(z, w) < \delta$ implies $d(f(z), f(w)) < \varepsilon/2$ for $z, w \in K$. By induction, choose $\delta_1 < \delta$ such that if $y_0, \ldots, y_m$ is a δ_1-chain in K then $d(f^m(y_0), y_m) < \delta$. Then $d(f^{m+1}(y_0), f(y_m)) < \varepsilon/2$ and $d(f(y_m), y_{m+1}) < \varepsilon/2$. It follows that $d(f^{m+1}(y_0), y_{m+1}) < \varepsilon$. By induction the proof is complete. ∎

THEOREM 4.A.5: The following statements about a compact invariant set are equivalent:

(1) S is an attracting set.
(2) S is asymptotically stable.

Proof: To show that an attracting set S is asymptotically stable, choose an attractor block N for S. By Theorem 2.B.1, the interior of N is contained in $W^s(S)$, hence $W^s(S)$ is open. To show that S is stable, let U be a neighborhood of S. There exists an integer m such that $f^j(N) \subset U$ for all $j \geq m$. Since f is continuous and S is invariant, each point x in S has a neighborhood V_x such that $f^j(V_x) \subset U$ for $0 \leq j \leq m$. Since S is compact, finitely many such neighborhoods cover S. Therefore, there is a neighborhood V of S such that $f^j(V) \subset U$ for $0 \leq j \leq m$. By the choice of m, $f^j(V \cap \text{int } N) \subset U$ for all $j \geq 0$.

We now show that a set S that is asymptotically stable is attracting. By Proposition 4.A.2 there exists a positive integer m and a compact neighborhood V of S which is contained in $W^s(S)$ such that V is an attractor block for f^m. By Proposition 4.A.3 there exists $\varepsilon > 0$ such that $\text{ch}(S, X, \varepsilon, f^m) \subset V$. By Proposition 4.A.4 there exists $\delta > 0$ such that if $y_0, \ldots, y_{m-1}, \ldots, y_{2m-1}, \ldots$ is a δ-chain, then $y_0, y_{m-1}, y_{2,n-1}, \ldots$ is an ε-chain for f^m. Since S is an invariant set, one can add an initial orbit segment of arbitrary length at the beginning of a chain starting in S. Therefore, $\text{ch}(S, X, \delta, f) \subset \text{ch}(S, X, \varepsilon, f^m)$. By Proposition 4.A.1, $N = \text{ch}(S, X, \delta, f)$ is an attractor block contained in V. Since N is contained in the stable set of S, Theorem 2.B.1 implies that S is the maximal invariant subset of N. Therefore, S is attracting. ∎

Definition: A nonempty invariant set A is a *quasi-attractor* if it is the intersection of a nested sequence of attracting sets.

PROPOSITION 4.A.6: An invariant set is a quasi attractor if and only if it is chain-stable.

Proof: This follows from Propositions 4.A.1 and 4.A.2. ∎

The next theorem asserts that attractor blocks persist when the map is perturbed.

THEOREM 4.A.7: Suppose that N is an attractor block for f, and let r equal the Euclidian distance between $f(N)$ and $X - N$. If g is a map which is close to f on N in the sense that $d(f(x), g(x)) < r$ for all $x \in N$, then $g(N)$ is contained in the interior of N and hence N is an attractor block for g.

Proof: Let x be a point of N. Then the Euclidian distance between $g(x)$ and $f(N)$ is less than r since the distance between $g(x)$ and $f(x)$ is less than r. Therefore, $g(x)$ does not belong to the closure of $X - N$. Thus $g(x)$ belongs to the interior of N. ∎

When one studies a continuous parametrized family of maps, attracting sets can be "continued" as the parameter varies. Suppose that f_λ is a parametrized family of maps, and suppose that A_{λ_0} is an attracting set for the map f_{λ_0}. Let N be an attractor block for A_{λ_0}. By Proposition 4.A.7, N is an attractor block for f_λ when λ is close to λ_0. One defines the *continuation* A_λ of A_{λ_0} by the equation $A_\lambda = \cap \ \{f_\lambda^n(N): n \geq 0)\}$. Naturally one wants to study how the topology and orbit structure of the sets A_λ changes as the parameter varies, and what prevents further continuation. As we will see in the next section, isolated invariant sets can also be continued.

Bifurcation theory is concerned with continuing invariant sets over a maximal parameter range, and asking what happens to prevent further continuation. The basic results in this theory concern continuation of periodic points. A bifurcation theory for more general invariant sets is not yet well developed.

B. Isolated Invariant Sets and Isolating Blocks

Attracting sets are examples of isolated invariant sets and attractor blocks are special isolating blocks. Isolated invariant sets occurring in parametrized families of maps can be continued using isolating blocks in the same way that attracting sets are continued using attractor blocks. A study of their continuations could form a natural part of bifurcation theory.

Basic sets arise as the intersection of nested families of isolating blocks. Orbits inside isolating blocks can be studied in terms of the topology of the blocks and qualitative properties of maps restricted to blocks. Since an isolating block for a map f is a stable geometric object, information obtained by its use will persist when the map f is perturbed.

Given a compact subset N of X, the following sets are important for our study:

$$W^s(N,f) = \cap \ \{f^k(N) : k \leq 0\} \quad \text{(the local stable set of } N\text{)}$$

$$\text{Inv}(N,f) = \text{the maximal invariant set contained in } N$$

Remarks: If an orbit segment is contained in an isolating block N, then from the definition, all but the first and last points of the segment belong to the interior of N. An alternate way of defining an isolating block is to say that it is a compact set N such that $f^{-1}(N) \cap N \cap f(N)$ is contained in the interior of N. When f is a homeomorphism, a useful way to verify that a set N is a block is to check that $N \cap f(N) \cap f^2(N)$ is contained in the interior of $f(N)$. Recall that the *exit set* E for N is the set $E = \{x \in N : f(x) \notin N\}$ and the *exit threshold set* e for N is the set $e = \{x \in N : f(x) \in \partial N\}$.

Example: Consider the map f of the real numbers defined by $f(x) = 2x$. Choose N to be the interval $[-1, 1]$. Then N is an isolating block for any map g sufficiently close to f. By inspection, $\text{Inv}(N,f) = \{0\}$. Using a connectedness argument, one can show that $\text{Inv}(N, g)$ is nonempty for g close to f (see Theorem 4.B.4).

Example: Consider the logistic map defined by $f(x) = 6x(1-x)$. Let $N = [-1.1, 1.1]$. N is an isolating block and one can show that $\text{Inv}(N,f)$ is a Cantor set.

Example: The rectangle with corners $ABCD$ pictured in Figure 3.3 is an isolating block for the horseshoe map discussed in Chapter 3, section E. The exit set for this block consists of the rectangle minus the strips C_0 and C_1 pictured in Figure 3.4. The maximal invariant set in the rectangle is a Cantor set, and the map restricted to this set is topologically conjugate to the shift on two symbols.

Example: A "branched" horseshoe map is a homeomorphism of the plane which has a hexagonal isolating block. The hexagon is mapped across itself as pictured in Figure 4.2. The exit set E from the hexagon is shaded and has four components. The exit threshold set e has five components as shown.

The branched horseshoe map is suggested by the geometry of the resonance zone for a type 2 trellis described in Chapter 3, section E. The block is formed by reshaping the trellis and slightly enlarging the hexagonal shaped region formed by alternating segments of unstable and stable manifolds pictured in Figure 4.3.

Example: A solid ball can be an isolating block for a map of R^3. Three different ways of mapping the ball through itself are pictured in Figure 4.4. In each of these cases, the exit set is exactly the same, the map on the exit set is the same, and the exit set has three components. Theorem 4.B.4 applies to

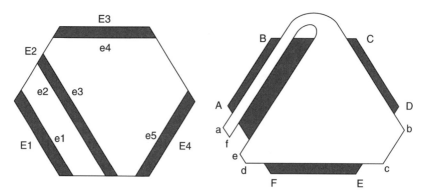

Figure 4.2. Hexagonal isolating block mapped across itself.

ISOLATED INVARIANT SETS AND ISOLATING BLOCKS   79

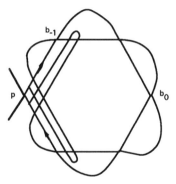

Figure 4.3. The trellis associated with the branched horseshoe map.

establish that the invariant set isolated by this block is nonempty for each of the three maps. One would like to know more about the invariant sets and the exit time decompositions of the ball in each case.

PROPOSITION 4.B.1: The intersection of two isolating blocks is an isolating block.

Proof: Let N_1 and N_2 be isolating blocks and let $N = N_1 \cap N_2$. Suppose that three points $x, f(x), f^2(x)$ on an orbit are contained in N. Then $f(x)$ is contained in the interior of both N_1 and N_2 since they are blocks. Hence $f(x)$ is contained in the interior of N. ∎

Exercise: The disjoint union of blocks is clearly a block. Construct an example where the union of two blocks which overlap is not a block.

When studying blocks and the invariant sets that they isolate, exit time functions as defined it Chapter 2, section F are very useful. These functions measure how long it takes for an orbit to leave a block. Since time is discrete and the state space is usually connected, one expects these functions to be discontinuous. They are discontinuous, but in the best possible way.

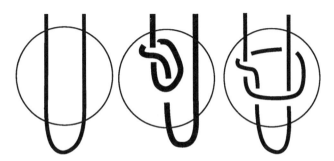

Figure 4.4. A solid ball mapped across itself in three different ways.

THEOREM 4.B.2: If N is an isolating block and x is a point of discontinuity of the exit time function t^+, then there is a neighborhood V of x in N such that for y in V, $t^+(y)$ is equal to either $t^+(x)$ or $t^+(x) - 1$, and $f^{t^+(x)-2}(x) \in e$, where e is the exit threshold set for N.

Proof: By continuity of the function f, there is a neighborhood U of x such that $t^+(y) \leq t^+(x)$ for $y \in U$. Thus, t^+ is continuous at points where $t^+ = 1$. Suppose that $2 \leq t^+(x) < \infty$. Since N is a block, the middle point of every three points along an orbit in N is contained in the interior of N. Therefore, the sequence $f(x), \ldots, f^k(x)$ belongs to the interior of N, where $k = t^+(x) - 2$. By continuity of the function f, there exists a neighborhood V of x such that the sets $f(V), \ldots, f^k(V)$ are contained in the interior of N. Therefore, $t^+(y) \geq t^+(x) - 1$ for $y \in V$. If $f^{k+1}(x)$ belongs to the interior of N, then t^+ is continuous at x. If $f^{k+1}(x)$ belongs to the boundary of N, then $f^k(x)$ belongs to the exit threshold set e, and t^+ may be discontinuous at x. If $t^+(x) = \infty$ then the orbit of x must be in the interior of N and nearby points must track this orbit for a long time. Thus, there is a neighborhood V of x where the exit time is large and therefore t^+ is continuous at x. ∎

PROPOSITION 4.B.3: Suppose that f maps open sets to open sets and suppose that $t^-(x) = k$. Then $t^-(x) \leq k$ on some neighborhood of x. If t^- is discontinuous at x, then $x = f^k(y)$ for some y belonging to the boundary of N.

The proof is left as an exercise.

The next theorem and its corollary illustrate how the exit time function can be used together with topological information to show that the set of orbits with infinite exit time is nonempty.

THEOREM 4.B.4: Suppose that N is an isolating block, and suppose that the exit set E is the disjoint union of two sets E_1 and E_2 which are open in N. Let $e_j = \text{cl}(E_j) \cap N - E_j$ and suppose that $f(e_j) \subset E_j$ for $j = 1, 2$. Define $U_k = \{x \in N - E : f^j(x) \in E_k \text{ where } j = t^+(x) - 1\}$. The the sets U_1 and U_2 are disjoint open subsets of $N - E$.

Proof: The sets U_1 and U_2 are disjoint because the sets E_1 and E_2 are disjoint. Let x be a point in U_1 such that the exit time function t^+ is constant in a neighborhood of x. Then by continuity of f, positive iterates of points in some smaller neighborhood of x track the orbit of x through the interior of N and hence exit through E_1. On the other hand, if the function t^+ it discontinuous at x then by Proposition 4.B.3 there is a neighborhood W of x such that every point in W exits in time $t^+(x)$ or in time $t^+(x) - 1$. Let $x^* = f^{t^+(x)-2}(x)$. The point x^* does not belong to E_2 since x belongs to U_1. The point x^* does not belong to the closure of E_2 because $f(e_2) \subset E_2$ and $f(x^*) \in E_1$. Choose a neighborhood V of x^* such that $V \cap E_2 = \emptyset$ and $f(V) \cap E_2 = \emptyset$. Thus, points in some small neighborhood of x which is contained in W track the orbit of x

through the interior of N and into V and hence exit through E_1. Therefore, U_1 is open. Similarly, U_2 is open. ■

COROLLARY 4.B.5: If C is a connected subset of N which intersects both E_1 and E_2, the C intersects $W^s(N,f)$.

Proof: If not, then $U_1 \cap C$ and $U_2 \cap C$ separate C into two pieces, but this is impossible because C is connected.

Example: Let $X = R^1$ and let $N = [-1, 1]$. N is a block for the map $f(x) = x^2 + \frac{1}{2}$, there is no invariant set contained in N, and yet the exit set from N consists of two disjoint half-open intervals $E_1 = (1/\sqrt{2}, 1]$ and $E_2 = [-1, -1/\sqrt{2})$. The reason the corollary does not apply is that $f(e_2) \subset E_1$.

Example: The theorem and corollary are false when N is not a block. Let $X = R^2$ and let $N = \{(x, y): |x| \le 2, -3 \le y \le 2\} - \{(x, y): x > 0$ and $y < 0\}$. Let $f(x, y) = (x + 1, y + 1)$. Then N is not a block, but the exit set has two disjoint components E_1 and E_2 which are pictured in Figure 4.5. The set U_1 dfined in the proof of Theorem 4.B.4 is not open. The orbit of the point $(-2, -2)$ exits through E_1, and intersects both the closure of E_2 and the boundary of N. The set N is connected, but N obviously contains no invariant set.

Exit times and exit time decompositions can be calculated for computer simulations of maps of Euclidian spaces. Points belonging to a two-dimensional slice of a set K can be colored according to their exit times and a rough picture of the set $W^s[K,f]$ can be displayed graphically (Easton et al. 1993). Indeed, this is essentially the way in which pictures of the Mandelbrot set are drawn.

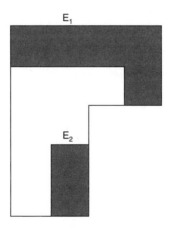

Figure 4.5. The set N and its exit sets. N is not an isolating block.

C. Constructing Isolating Blocks

Often one finds an isolating block based on familiarity with the dynamics generated by a given map. However, Theorem 4.C.1 gives a systematic method for constructing blocks using epsilon-chains. The set of chains that start and end in an isolated invariant set S is a block. If the map is invertible, it may be possible to use a computer to construct an isolating block. The block could be obtained as the intersection of the set of chains starting in S with the set of chains for the inverse of the map starting in S. In the case of an attracting set, an attractor block could be constructed just using forward chains.

THEOREM 4.C.1 (The Block Construction Theorem):

(1) If K is a compact subset of X and if S is a subset of K, and if $N = \text{ch}(S, S, K, \varepsilon)$ is contained in the interior of K, then N is an isolating block.
(2) If S is an isolated invariant set and K is an isolating neighborhood for S, then there exists $\varepsilon > 0$ such that $\text{ch}(S, S, K, \varepsilon) \subset K$. Further S is the maximal invariant subset of $\text{ch}(S, S, K, \varepsilon)$.

Proof: To show that N is a block, assume for the purpose of contradiction that $x, f(x)$, and $f^2(x)$ belong to N but that $f(x)$ is not in the interior of N. By Lemma 2.D.1, we can choose neighborhoods U, V, and W of $x, f(x)$, and $f^2(x)$, respectively, which are contained in the interior of K and are chosen so that any selection of three points, one in each set, forms an epsilon-chain. Choose an epsilon-chain in K from S to x, and an epsilon-chain in K from $f^2(x)$ to S. Since N is contained in the interior of K, there is a point z in V which is not in N. Then the chain obtained by concatenating the chain from S to x with z and then with the chain from $f^2(x)$ to S is an epsilon-chain in K from S to S through the point z. This implies that z belongs to N, which is a contradiction. This proves (1).

To prove (2), let $R = \text{ch}(S, S, K)$. Since S is the maximal invariant set in K, by Proposition 2.D.3, R is contained in S. Take a sequence of epsilons converging to zero. Then the corresponding sets $\text{ch}(S, S, K, \varepsilon)$ converge to R and therefore are contained in the iterior of K for sufficiently small epsilon. Since S is maximal in K, it also maximal in $\text{ch}(S, S, K, \varepsilon)$. ∎

THEOREM 4.C.2 (Stability of Blocks): If N is an isolating block for f, then there exists $\varepsilon > 0$ such that N is an isolating block for any map g of X which satisfies the condition:

$$\sup\{d(f(x), g(x)): x \in N\} < \varepsilon.$$

Proof: Suppose the theorem is false. Then there exists a sequence of functions g_n converging to f on N in the sense that

$$\sup\{d(f(x), g_n(x)): x \in N\} < 1/n$$

such that N is not an isolating block for g_n. Hence, there exists a sequence of points x_n such that $x_n, g_n(x_n), g_n^2(x_n) \in N$ and $g_n(x_n) \in \partial N$. Since N is compact we may assume that the points x_n converge to a point x in N. It follows that $g_n(x_n) \to f(x)$ and $g_n^2(x_n) \to f^2(x)$. Since $g_n(x_n) \in \partial N$ and $g_n^2(x_n) \in N$, it follows that $f(x) \in \partial N$ and $f^2(x) \in N$. This contradicts the fact that N is an isolating block for f. Thus the theorem is true. ∎

D. Basic Sets

Theorem 4.D.1 shows that basic sets are contained in nested sequences of isolating blocks. Thus, they cannot suddenly grow in diameter or explode when the map is perturbed. However, they may implode or vanish when the map is perturbed unless the way the block is mapped across itself forces a nonempty invariant set to exist. Thus, the directed graph associated with f is not entirely a stable object, nor is the Conley decomposition entirely stable. We will come back to study the stability of these objects further in Chapter 5.

THEOREM 4.D.1: If K is an isolating region, and if S is a K-basic set, then $S = \cap \{\text{ch}(S, S, K, \varepsilon): \varepsilon > 0\}$. Hence every K-basic set is equal to the intersection of a family of isolating blocks.

Proof: By Proposition 2.D.3, $\text{ch}(S, S, K)$ is an invariant set. Since K is an isolating region, $\text{ch}(S, S, K)$ is contained in the interior of K. Since S is a basic set, $\text{ch}(S, S, K) = S$. Since $\text{ch}(S, S, K) = \cap \{\text{ch}(S, S, K, \varepsilon): \varepsilon > 0\}$, it follows that, for all ε sufficiently small, $\text{ch}(S, S, K, \varepsilon)$ must be contained in the interior of K.

By Theorem 4.D.1, $\text{ch}(S, S, K, \varepsilon)$ is an isolating block for sufficiently small ε, and S is equal to the intersection of the isolating blocks $\text{ch}(S, S, K, \varepsilon)$. ∎

E. Symbolic Dynamics

Symbolic dynamics was used to analyze orbits for the horseshoe map. Similarly, it can be used to study orbits trapped inside an isolating block.

Definition: The *symbol space* Σ^+ on $m+1$ symbols is defined to be the set of all infinite sequences (or functions)

$$s: Z^+ \to \{0, 1, \ldots, m\}$$

The symbol space is a metric space with metric defined by the formula

$$d(z, t) = \sum_{j=0}^{\infty} 2^{-j} |s(j) - t(j)|$$

The *shift automorphism* $\sigma \colon \Sigma^+ \to \Sigma^+$ is defined by the formula

$$\sigma(s)(j) = s(j+1)$$

Certain invariant sets for the shift automorphism are important since they occur freuqently in applications. These are called the subshifts of finite type. Suppose that M is an irreducible $m \times m$ matrix with all entries 0 or 1. *Irreducible* means that for each (i, j) there is a power of M which has positive entry in the (i, j) position. Define the *subshift symbol space* $\Sigma^+[M]$ by

$$\Sigma^+[M] = \{s \in \Sigma^+ : \text{for each } i, \text{ the entry of } M \text{ in position } s(i)\ s(i+1) \text{ is } 1\}$$

The set $\Sigma^+[M]$ is an invariant set for σ and the restriction of σ to $\Sigma^+[M]$ is called a *subshift of finite type*.

Suppose that N is an isolating block for a map f of a metric space X. Recall that $N[r] = \{x \in N : t^+(x) \geq r\}$. Suppose that $N[r]$ is the disjoint union of compact sets $C_0, \ldots, C_m$. For each point x belonging to $S(N)$ define an *itinerary* $s[x, f]$ to be the sequence

$$s[x, f](j) = k \text{ if and only if } f^j(x) \in C_k$$

Define the *itinerary map* $\tau \colon S(N) \to \Sigma^+$ by the formula $\tau(x) = s[x, f]$. From these definitions one has a commuting diagram:

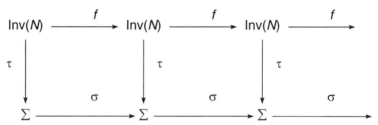

PROPOSITION 4.E.1: If N is an isolating block and $N[r]$ is the disjoint union of compact sets $C_0, \ldots, C_m$, then the itinerary map is continuous.

Proof: The fact that $N[r]$ is an isolating block ensures that orbits march through the interiors of the compact sets. Thus, finite-length segments of nearby initial states have the same itineraries. This makes τ continuous. ∎

More analysis is needed to characterize the range of the itinerary map. One may form a matrix M by setting $M_{ij} = 1$ if $f(C_i)$ intersects C_j and $M_{ij} = 0$ if $f(C_i)$ does not intersect C_j. If there are orbits in $S^*(N)$ which go from C_i to C_j

for any i and j, then the admissable itineraries are contained in the subshift symbol space $\Sigma^+[M]$. In some cases of interest one can show that $\Sigma^+[M]$ is the range of the itinerary map, and further that the itinerary map is a topological conjugacy,

If the range of the itinerary map τ is $\Sigma^+[M]$, then τ is a semiconjugacy between f and the shift σ on $\Sigma^+[M]$. It follows from a standard result (Bowen, 1975) that the topological entropy of f is greater than or equal to the topological entropy of σ.

Exercise: Compute the topological entropies of general subshifts of finite type.

F. Filtrations of Isolated Invariant Sets

Filtrations factor invariant ses into Morse sets. The "prime factors" are basic sets. However, an invariant set may contain infinitely many basic sets, and thus may not factor into finitely many primes.

Definition: A *filtration* of a compact confining set S for a map f of a metric space X is a finite collection $A_0 \subset A_1 \subset \cdots \subset A_r$ of compact sets such that

1. A_0 is the empty set, and A_r is equal to S.
2. $f(A_j)$ is contained in the interior (relative to S) of A_j.
3. A_j is contained in the interior (relative to S) of A_{j+1}.

Let M_k denote the maximal invariant set contained in $A_k - A_{k-1}$. An invariant set which arises in this way is called a *Morse set*. A *short filtration* is a filtration with $r = 2$.

Example: Let f be the logistic map $f(x) = 2x(1 - x)$. The interval $[0, 0.5]$ is a confining set, and the sets $A_2 = [0, 0.5]$, $A_1 = [0.25, 0.5]$, $A_0 = \emptyset$ form a filtration with Morse sets $M_2 = \{0\}$ and $M_1 = \{0.5\}$.

THEOREM 4.F.1: If N is an isolating block and B is an N-bsic set, then there is no nontrivial filtration of B.

Proof: Basic sets are chain transitive and having a filtration would prevent chain transitivity. Therefore, basic sets cannot be factored. ∎

Next we show that resticting the dynamics to a basic set does not produce new basic sets.

THEOREM 4.F.2: If N is an isolating block and B is an N-basic set, then B is a basic set for f restricted to B.

Proof: Suppose that B is an N-basic set. Since B is invariant, it is contained in the interior of N. Given two points of B and a positive epsilon, there is an epsilon-chain in N from each to the other. To show that B is a basic set within itself one must construct an epsilon-chain in B between any pair of points in B. Choose $0 < \delta < \varepsilon/3$ such that, if x and y belong to N, then $d(x, y) < \delta$ implies that $d(f(x), f(y)) < \varepsilon/3$. Choose δ smaller if necessary so that any δ-chain from B to B is contained in an $\varepsilon/3$ neighborhood of B. Given points x and y in B, choose a δ-chain $y_1, \ldots, y_r$ in N from x to y. For $1 < j < r$ choose a point z_j in B with $d(y_j, z_j) < \varepsilon/3$. By the choice of δ, the set of points $x, z_1, \ldots, z_{r-1}, y$ is an ε-chain in B from x to y. ∎

THEOREM 4.F.3: If S is an isolated invariant set which is not a basic set, then S has a nontrivial filtration.

Proof: Since S is not a basic set, there exists $\varepsilon > 0$ and two points x and y belonging to S such that there is no ε-chain from x to y in S. Let F_1 be the set of all ε-chains in S which start at x. Let F_0 be the empty set and let $F_2 = S$. Then $\{F_1, F_1, F_2\}$ is the desired filtration. ∎

G. Stacks of Isolating Blocks

A basic problem-solving strategy involves analysis and synthesis: breaking a big problem into smaller ones, solving these, and putting the parts back together. As part of this strategy, it is natural to want to factor a "big" isolating block, into smaller blocks. Factoring a block will mean representing it as the union of a stack of blocks.

Definition: A *stack of blocks* is a finite collection of blocks $N_1, \ldots, N_r$ such that

1. The interiors of the blocks are disjoint.
2. $N = N_1 \cup \cdots \cup N_r$ is a block with exit set E.
3. If x is in the exit set of N_j, then either x belongs to E of $f(x)$ belongs to N_{j-1}.

THEOREM 4.G.1: Given an isolating block N and a short filtration $A_0 \subset A_1 \subset A_2$ of $W^s(N, f)$, there exists a short stack of blocks N_1, N_2 which factors N, such than $\text{Inv}(N_2) = M_2$ and $\text{Inv}(N_1) = M_1$, where M_1 and M_2 are the Morse sets associated with the filtration.

Proof: Choose ε sufficiently small so that all ε-chains in $W^s(N, f)$ which start on the Morse set M_1 are contained in A_1. Define $N_1 = \text{ch}(M_1, N, \varepsilon)$ and define $N_2 = N - \text{int } N_1$. The sets N_1 and N_2 will form the desired stack of blocks.

To show that N_1 is a block, suppose that x belongs to the intersection of the boundary of N_1 with the interior of N. If $f(x)$ belongs to the boundary of N, then since N is a block $f^2(x)$ does not belong to N_1. If $f(x)$ belongs to the interior of N, we will show that it also belongs to the interior of N_1. If not, then by Lemma 2.D.1 there exist neighborhoods U of x and V of $f(x)$ contained in the interior of N such that any point in U with any point in V forms an ε-chain. If $f(x)$ belongs to the boundary of N_1, then some point q in V does not belong to N_1. However, there is a chain to some point in U and hence there is a chain to the point q in V. This is a contradiction. Therefore, $f(x)$ belongs to the interior of N_1, and it follows that N_1 is a block.

To show that N_2 is a block, suppose that y belongs to N_2. If $f(x)$ belongs to the boundary of N, then since N is a block, $f^2(x)$ does not belong to N_2. If $f(x)$ belongs to the boundary of N_2 and is in the interior of N, then $f(x)$ also belongs to the boundary of N_1. By the preceding argument, $f^2(x)$ belongs to the interior of N_1. Thus N_2 is a block.

N is the union of the blocks N_1 and N_2. Further, any point which exits from N_2 either exits N or enters N_1. Thus N_1 and N_2 form a stack of blocks.

Since there is no ε-chain from M_1 to M_2 in N (for ε sufficiently small), M_2 must be the maximal invariant set contained in N_2. ∎

COROLLARY 4.G.2: If the filtration is $A_0 \subset A_1 \subset \cdots \subset A_r$, then there exists a stack of blocks $N_1, \ldots, N_r$ such that $N = N_1 \cup \cdots \cup N_r$ and $\mathrm{Inv}(N_j) = M_j$.

Proof: Split N into two blocks using Theorem 4.G.1 so that the second block contains the Morse set M_r. Continue using Theorem 4.G.1 to split the first block until the desired stack is created. ∎

Conclusion: If the stable set $W^s[N,f]$ of a block N can be "factored" into Morse sets by a filtration, then N can be factored into a stack of blocks which isolate the Morse sets.

Remark: A stack of blocks reflects the "gradient-like" structure of the dynamics. For volume-preserving maps the dynamics is not globally gradient-like. However, a network of blocks may exist. The difference between a network and a stack of blocks is that orbits can reenter a block in a network whereas in a stack they must progress from the top to the bottom.

Definition: A *network of blocks* is a finite collection of blocks $N_1, \ldots, N_r$ such that

1. The interiors of the blocks are disjoint.
2. $N = N_1 \cup \cdots \cup N_r$ is a block with exit set E.
3. If x is in the exit set of N_j, then either x belongs to E or $f(x)$ blongs to N_k for some k.

88   GEOMETRIC METHODS FOR DISCRETE DYNAMICAL SYSTEMS

Research Question: Can one "tile" the phase space of a volume-preserving map with a network of blocks in such a way that useful information about the dynamics can be obtained?

H. Calculating Directed Graphs

For an isolated invariant set S it may not be possible in practice to determine all the S-basic sets. Rather, one may have only partial information about invariant sets contained in S and their relation to each other. Perhaps one can find an isolating block N for S and a disjoint collection of blocks N_j each contained in N. One wants to record information about how orbits in the big block N connect the small blocks to each other or connect the small blocks to various components of the exit set. As in Chapter 2, this is done by forming a "directed graph."

Definition: An *array of blocks* is a collection of blocks $\{N, B_1, \ldots, B_r\}$ such that the blocks B_j are disjoint, and each is contained in N. The *directed graph* for this array is constructed as follows: For each small block and for each component E_j of the exit set E of N, assign a *block vertex* or an *exit vertex*. A directed edge joins a given pair of block vertices $v(B)$ and $v(B')$ if and only if there is a full orbit $\{x_j\}$ contained in N such that x_j is contained in B except for at most finitely many negative j's, and x_j is contained in B' except for at most finitely many positive j's. Assign an edge from a block vertex $v(B)$ to an exit vertex $v(E_k)$ if there is an full orbit $\{x_j\}$ such that x_j is contained in B except for at most finitely many negative j's, and which exits N through the component E_k of the exit set.

Given an array of blocks $\{N, B_1, \ldots, B_r\}$, one can partition N in the following way: Let L denote the set of points in N which exit N in a finite time. The orbit of any point of $N - L$ enters and remains in the block B_j which contains its omega limit set. Thus, the stable sets $W^s(B_j)$ together with L form a partition of N. This partition is similar to the partition determined by the Conley decomposition theorem of Chapter 2.

Given an array of blocks, one wants to use the topology of these blocks together with rough information about the way the map f acts to investigate the orbit structure in N. A goal would be to compute the associated directed graph and the partition of N. A local analysis of the dynamics inside each small block would complete a satisfying study of the dynamics in N. While in theory this analysis is possible, it has not yet been applied to a practical problem.

Example: Suppose that g is a diffeomorphism of the plane which has four fixed points: a source, a sink, and two saddles. These may be arranged at the corners of a square as pictured in Figure 4.6. Suppose that f is a map

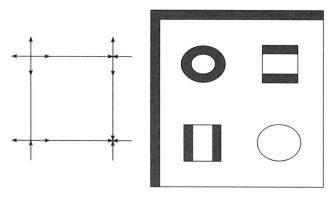

Figure 4.6. A diffeomorphism of the plane with four fixed points. Isolating blocks containing the fixed points are shown on the right. Their exit sets are shaded.

close to g. Suppose that for f, one knows that there is an isolating block N in the shape of a square together with four blocks B_1, B_2, B_3, B_4 as pictured. The exit sets for the five blocks are shaded. The problem is to say as much as possible about the orbit structure of g in N.

Example: The N-directed graph associated with the blocks pictured in Figure 4.6 has four block vertices and one exit vertex. Without further analysis one does not know how to draw the directed edges. However, if the dynamics is as pictured on the right side of Figure 4.6 then the N-directed graph is pictured in Figure 4.7.

Question: Given as an isolating block N, when should one stop looking for new small blocks inside N? Suppose that one has found the big block and the left two small blocks in Figure 4.6. How does one know that there are more N-basic sets which are not contained in the two small blocks? The Conley index, which is discussed in Chapter 5, is a tool designed to help answer this question.

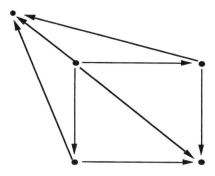

Figure 4.7. The directed graph associated with the blocks pictured in Figure 4.6.

Question: Given an array of blocks, when should one draw an edge in the directed graph? If N can be factored into a stack of blocks so that the small blocks are contained inside individual blocks in the stack, then the edges can only be drawn "downward" relative to the stack. The subsequent discussion is directed toward proving Theorem 4.H.3 which allows one to draw an edge.

Definition: Given an isolated invariant set S, and given a disjoint pair F, G of invariant sets contained in S, define x to be a *heteroclinic point* from F to G in S if there exists a full orbit $\{x_k\}$ contained in S with $x_0 = x$ such that $d(x_k, F) \to 0$ as $k \to -\infty$ and $d(x_k, G) \to 0$ as $k \to \infty$. Define $\text{Het}(F, G)$ to be the set of all such heteroclinic points.

Example: When the sets S, F, G are isolated invariant sets, the set $F \cup G \cup \text{Het}(F, G)$ is not necessarily an isolated invariant set. To see this, let $S = [0, 1] \times [0, 1]$ and construct a map f with the following properties: The fixed points of f consist exactly of the points $(0, 0.5)$ and $(1, t)$ for $0 \le t \le 1$. Require f to move all other points in the square horizontally to the right. Then the sets $F = \{(0, 0.5)\}$ and $G = \{(1, t): 0 \le t \le 1\}$ are isolated invariant sets, but the set $F \cup G \cup \text{Het}(F, G)$ is not isolated because there are orbits not in this set which start close to F and remain uniformly close to it (their omega limit sets are in G).

Definition: Given an isolated invariant set S, and given a disjoint pair F, G of isolated invariant sets contained in S, a *connection* from F to G in S is an isolated invariant set Λ with $(F \cup G) \subset \Lambda \subset S$ such that if $x \in \Lambda - F$ then $f^n(x) \to G$ as $n \to \infty$ and, furthermore, G is an attracting set in Λ.

Example: Note that the last requirement that G be an attracting set is not redundant. Consider a space Λ consisting of the disjoint union of a fixed point q and a circle. Assume that the circle is mapped to itself as in the first example of section 4.A. Let $F = q$ and let G be the unique fixed point on the circle. The set Λ has all the properties of a connection except that G is not attracting.

Each of the sets F, G, Λ has a continuation when the map f belongs to a parametrized family of maps. One wants to know when a connection persists. This is a topic of current research, and Theorem 4.H.3 below shows that some connections persist.

THEOREM 4.H.1: Suppose that Λ is a connection from F to G. Then there exists a filtration $A_0 \subset A_1 \subset A_2$ with Morse sets $M_1 = G$ and $M_2 = F$.

Proof: We consider Λ as the phase space on which we are working and forget the larger context. It is sufficient to construct an attractor block A_1 containing G. By Theorem 4.A.5 such a block can be constructed since the set G is an attracting set. Since we have assumed that the orbits of all points not in F converge to G, the Morse set M_2 is equal to F. ∎

THEOREM 4.H.2: Suppose that Λ is a connection from F to G. Then there exists a short stack of blocks (N_1, N_2) such that N_2 isolates F and N_1 isolates G and $N_1 \cup N_2$ isolates Λ.

Proof: By Theorem 4.H.1, there is a filtration of Λ with Morse sets F and G. The short stack is constructed using Theorem 4.G.1.

The next theorem could be used to justify drawing a directed edge between two vertices.

THEOREM 4.H.3: Suppose a short stack of blocks N_1, N_2 is given. Suppose that the exit set E_1 for N_1 has exactly two components. Suppose that $W^u(N_2) \cap N_1$ has a connected subset which intersects both components of E_1. Then $\text{Het}(S^*(N_2), S^*(N_1))$ is nonempty.

Proof: By Corollary 4.B.5, the connected subset of $W^u(N_2) \cap N_1$ which intersects both components of E_1 contains a point belonging to $W^s(N_1)$. This is the desired heteroclinic point. ∎

Example: Theorem 4.H.3 may seem artificial, but this example illustrates its use. Cartesian products of dynamical systems are the analog of decoupled systems of differential equations. This example is built by weakly coupling a product of maps. Define $g: R^2 \to R^2; (x, y) \to (x^3, 2y)$ and define

$$N_1 = [-\tfrac{1}{2}, \tfrac{1}{2}] \times [-1, 1], \qquad N_2 = [-\tfrac{3}{2}, -\tfrac{1}{2}] \times [-1, 1]$$

Then the pair N_1, N_2 is a short stack of blocks for f. N_2 is a repeller block, and therefore $W^u(N_2) \cap N_1 = N_2 \cap N_1 = -\tfrac{1}{2} \times [-1, 1]$. The exit set for N_1 has two components, and hence Theorem 4.H.3 applies to any map f which is a sufficiently small perturbation of g.

Problem: Define $g: R^3 \to R^3; (x, y, z) \to (x^3, 2y, z/2)$, and define

$$N_1 = [-\tfrac{1}{2}, \tfrac{1}{2}] \times [-1, 1] \times [-1, 1], \qquad N_2 = [-\tfrac{3}{2}, -\tfrac{1}{2}] \times [-1, 1] \times [-1, 1]$$

Show that any map f which is sufficiently small perturbation of g has an orbit heteroclinic between the (nonempty) invariant sets contained in N_1 and N_2. Show that the set $W^u(N_2)$ has a connected subset which intersects both of the disjoint components of the exit set of N_1.

So far, point set topology has supplied the concepts and tools sufficient for our analysis, and we have avoided using algebraic topology. In Chapter 5 we use algebraic topology to generalize Theorem 4.H.3 and to prove that heteroclinic orbits exist.

I. Further Reading

The essential reference for Conley's approach to dynamics is his monograph (Conley, 1978). Some of the recent research concerning isolated invariant sets appears in the Conley memorial volume (Herman et al., 1988). McGehee (1988) used epsilon-chains to construct attractor blocks and suggested to me that they might be used to construct isolating blocks.

The Conley index for discrete dynamical systems was developed by Mrozek (1990) using index pairs rather than isolating blocks. Extensions of the Conley index to dynamics in infinite-dimensional phase spaces is discussed by Rybakowski (1987). Smoller (1983) gives the theory of the Conley index for flows with emphasis on heteroclinic connections between invariant sets and applications to partial differential equations. Further development of the index for flows is presented in Kurland (1996).

5

The Conley Index

In this chapter, we investigate properties of the set of orbits trapped by an isolating block. One can restrict the dynamics to a block by collapsing its exit set. The quotient space derived from the block is called the index space. A derived dynamical system called the index map is defined on this space and has the collapsed exit set as a fixed point. The index space together with the index map is called the *Conley index of the block*. An induced map on the homology groups of the index space is the index homomorphism. If the index homomorphism is nontrivial, then so is the set of orbits trapped by the block. The Conley index for an isolated invariant set is defined by taking the direct limit of the indices of a sequence of isolating blocks which converge to the set.

A. The Conley Index of an Isolating Block

Definition: If (X, d) is a metric space, and R is an equivalence relation on X, then the set of equivalence classes of R is denoted by X/R. There is a projection of X onto X/R defined by $P(x) = [x]$, where $[x]$ denotes the equivalence class of x. The *quotient topology* on X/R is defined exactly to make the projection map continuous. Thus a subset U of X/R is defined to be open if and only if $P^{-1}(U)$ is open in X. The set X/R with the quotient topology is called the *quotient space* of the equivalence relation R.

Definition: Suppose that N is an isolating block for a map f of the metric space X with nonempty exit set E and exit threshold set e. Define an equivalence relation on N whose equivalence classes consist of single points for each point not belonging to $E \cup e$ and define all points of $E \cup e$ to be one equivalence class denoted by $*$. Denote the quotient space associated with this equivalence relation by $N^{\#}$. This space is called the *index space* of the block. The *index map* $f^{\#}: N^{\#} \to N^{\#}$ is defined by

$$f^{\#}(x) = \begin{cases} * & \text{if } x \in E \cup e \\ * & \text{if } f(x) \in E \cup e \\ f(x) & \text{otherwise} \end{cases}$$

The *Conley index of the block* is the pair $(N^{\#}, f^{\#})$.

Remark: It seems to be necessary to first collapse the entire exit set before restricting the dynamics to the block. Otherwise the restricted map may not be continuous. However, in some cases it is possible to separately collapse components or groups of components of the exit set.

Example: Let f be the logistic map on the real line given by $f(x) = 9x(1-x)$, and let N be the closed interval $[-1, 2]$. Then N is an isolating block for f. The exit set E consists of the intervals $[-1, a]$, (b, c), $(d, 2]$, where $f(a) = f(c) = -1$ and $f(b) = f(c) = 2$. The index space $N^{\#}$ is topologically equivalent to a figure-of-eight, or to two circles with exactly one point in common. The function f is monotonically increasing on the interval $[a, b]$, and maps this interval onto N. The function f is monotonically decreasing on the interval $[c, d]$ and maps this interval onto N. The set $N[2]$ of points in N with exit time greater than or equal to 2 consists of four closed intervals. The index space of $N[2]$ is a clover-leaf, or the space of four circular curves having exactly one point in common. It is important to note that the blocks $N[1]$ and $N[2]$ both isolate the same invariant set, and yet have different index spaces. Thus, it is necessary to go beyond the index spaces and index maps of blocks in order to assign an "index" to the invariant set isolated by a block.

THEOREM 5.A.1: The index space $N^{\#}$ is a compact metric space, and the index map $f^{\#}$ is continuous. The base point $*$ of $N^{\#}$ is an attracting fixed point. The complement of its basin of attraction consists of all points which never exit from N.

Proof: $N^{\#}$ is compact since it is the continuous image under the quotient projection of the compact space N. It is also Hausdorff, since N is Huasdorff. It is a theorem in topology that compact Hausdorff spaces have a metric compatible with their topology. The base point $*$ is attracting since the open set $E[2] \cup *$ is a neighborhood of $*$ which maps to $*$. The basin of the fixed point $*$ is the projection of $N - W^s[N]$, and hence $W^s(N)$ is the complement of the basin of attraction. ∎

The stable set $W^s(N)$ of a block N is the intersection of the blocks $N[k]$ over all positive integers k. Each of these blocks has an associated index space and index map. The relationship between these spaces and maps can be used to gain information about the set $W^s(N)$. The notation $N^{\#}[k]$ is used to denote the index space of $N[k]$.

Definitions: There is a "projection" map $\gamma_k: N^\#[k] \to N^\#[k+1]$ defined by the formula

$$\gamma_k(x) = x \text{ if } x \in N[k+1] - e[k+1] \quad \text{and} \quad \gamma_k(x) = * \text{ otherwise}$$

The asterisk $*$ denotes the base point of the index space $N^\#[k+1]$.
The map $\phi_k: N^\#[k] \to N^\#[k-1]$ induced by f is defined by the formula

$$\phi_k(x) = f(x) \text{ if } f(x) \notin e[k-1] \quad \text{and} \quad \phi_k(x) = * \text{ otherwise}$$

Note that the composition of maps $\gamma_k \phi_{k+1}$ is the index map for the block $N[k]$.

THEOREM 5.A.2: The kth power of the index map $f^\#$ of a block N is related to the maps defined above by the formula $(f^\#)^k = \phi_2 \cdots \phi_{k+1} \gamma_k \cdots \gamma_1$.

Proof: Consider the diagram pictured in Figure 5.1. Since it locally commutes, following arrows from one place to another is independent of the path. Thus, for example, $(f^\#)^3 = \phi_2 \phi_3 \phi_4 \gamma_2 \gamma_2 \gamma_1$. The general result follows from tracing arrows around a larger diagram. ∎

The spaces $N[k]$ for a block N are changing as k increases and are not easy to calculate. If they could be calculated, one would know a great deal about the dynamics of a map f inside the block N. However, qualitative properties of the index map of the block can be calculated. For the following argument the reader is assumed to know about the fundamental groups of spaces and the homomorphisms of these groups associated with maps between spaces. A brief discussion of fundamental groups is included in Appendix E.

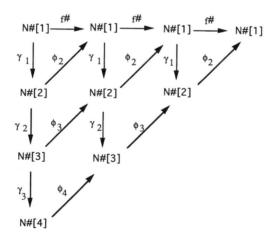

Figure 5.1. Diagram of index spaces showing projection and quotient maps between the spaces.

PROPOSITION 5.A.3: The index map $f^{\#}$ of a block N induces a homomorphism $f\#$ of the fundamental group of the index space. Suppose that there is an element z of $\pi_1(N^{\#}[1])$ such that $f_{\#}^k(z) \neq 1$ for each k. Then the stable set $W^s(N)$ of the block in not empty.

Proof: It is sufficient to show that the sets $N[k]$ are nonempty for each k. The homomorphism of fundamental groups associated with a composition of maps is equal to the composition of the homomorphisms associated with each map. The projection homomorphism $(\gamma_k \ldots \gamma_1)_{\#}(z)$ of z into the fundamental group of $N^{\#}[k]$ is nontrivial because, by Theorem 5.A.2,

$$(f^{\#})^k(z) = \phi_2 \ldots \phi_{k+1}\, \gamma_k \ldots \gamma_1(z)$$

and by hypothesis $f_{\#}^k(z) \neq 1$. Since the fundamental group of $N^{\#}[k]$ is nontrivial, the space $N[k]$ is nonempty. Hence $W^s[N]$ is nonempty because it is the intersection of the nested sequence of compact sets $N[k]$. ∎

Example: Let f be a map of the plane, and suppose that N is a square centered at the origin which is an isolating block for f. Suppose that the exit set E of N consists of two vertical strips which include the vertical sides of the square (as is the case for the linear map $(x, y) \to (2x, y/2)$. Then the index space $N^{\#}$ with respect to f is homeomorphic to a pinched annulus A. The set A is the region bounded by the circle of radius 2 centered at $(0, 2)$ and the circle of radius 1 centered at $(0, 1)$. The fundamental group of A is the integers. If the index homomorphism is nontrivial, then by Proposition 5.A.3 the invariant set isolated by N is nonempty.

The uses and properties of groups associated in natural ways with topological spaces is the subject of algebraic topology, which is very briefly reviewed in Appendix E. A difficulty in using fundamental groups and higher homotopy groups is that they are generally hard to compute. Homology groups are fairly easy to compute, and they will be used to define an index in section C.

B. Continuation of Isolated Invariant Sets

Definitions: Let $g_\lambda(x)$ be a one-parameter family of maps for $0 \leq \lambda \leq \alpha$, and suppose that N is an isolating block for each map in this family. Then the invariant set I_α of the map g_α which is isolated by N is called the *local continuation* of the invariant set I_0 of the map g_0 which is isolated by N. An isolated invariant set I_0 for the map g_0 has a *continuation* I_λ with respect to λ if there is a finite sequence of local continuations joining I_0 and I_λ.

Let $N_\lambda^{\#}$ denote the index space of N with respect to the map g_λ and let $g_\lambda^{\#}$ denote the associated index map. Since the exit set from the block N can change as the parameter changes, the index spaces may not all be homeomorphic.

THE CONLEY INDEX   97

Example: Consider the logistic maps $L(x) = \lambda x(1 - x)$. For λ near 6, the interval $N = [-1, 1.5]$ is an isolating block. The index spaces $N_\lambda^\#$ change from being circles to being the wedge of two circles at the parameter value $\lambda = 6$.

In the context of differentiable maps on smooth manifolds, one can construct isolating blocks which are manifolds with their boundaries mapped transversally across themselves. Then by techniques of differential topology, one can establish that the index spaces are diffemorphic for C^1 cloes maps. This approach is not used here because the technical results would take a considerable digression to develop. Instead, our goal is to relate the index spaces $N_\lambda^\#$ and index maps $g_\lambda^\#$ to each other for λ close to 0 by projection maps. To do this we use a general result about "nested" quotient spaces.

The term pair of spaces (A, B) indicates that B is always a subset of A.

PROPOSITION 5.B.1: Suppose that (A, B) and (C, D) are pairs of compact spaces, and further suppose that $C \subset A$ and $B \cap (C - D) = \varnothing$. Then there is a continuous projection map $\pi: A/B \to C/D$ defined by the formula

$$\pi(x) =: \begin{cases} x \text{ if } x \in C - D \\ * \text{ otherwise} \end{cases}$$

The proof is left as an exercise.

Recall that for a block N, the sets $N[k]$ are also isolating blocks, and each has its associated index space $N^\#[k]$. Using Proposition 5.B.1, there are projection maps $\gamma_k^j: N^\#[k] \to N^\#[j]$ for $k < j$. The index map of $N_\lambda^\#[k]$ is denoted by $g_\lambda^\#$.

THEOREM 5.B.2: Suppose that N is an isolating block for a one-parameter family of maps $g_\lambda(x)$. Then for λ sufficiently small, there are projection maps $\pi: N_0^\#[1] \to N_\lambda^\#[3]$ and $\tilde{\pi}: N_\lambda^\#[1] \to N_0^\#[3]$. The diamgrams below almost commute in the sense that $g\#_\lambda \pi(x) \to \pi g_0^\#(x)$ as $\lambda \to 0$, and $g_0^\# \tilde{\pi}(y) \to \tilde{\pi} g_\lambda^\#(y)$ as $\lambda \to 0$ for all $x \in N_0^\#[1]$ and $y \in N_\lambda^\#[1]$.

$$\begin{array}{ccc} N_0^\#[1] \xrightarrow{g_0^\#} N_0^\#[1] & & N_\lambda^\#[1] \xrightarrow{g_\lambda^\#} N_\lambda^\#[1] \\ \pi \downarrow \quad \downarrow \pi & & \tilde{\pi} \downarrow \quad \downarrow \tilde{\pi} \\ N_\lambda^\#[3] \xrightarrow{g_\lambda^\#} N_\lambda^\#[3] & & N_0^\#[3] \xrightarrow{g_0^\#} N_0^\#[3] \end{array}$$

Proof: It is sufficient to show that $N_\lambda[3] \subset N_0[1] - E_0[1] \cup e_0[1]$, and then use Proposition 5.B.1 to construct $\pi: N_0^\#[1] \to N_\lambda^\#[3]$. Each point of the set $E_0[1] \cup e_0[1]$ maps strictly outside N by the first or second iterate of g_0. Thus, there is a positive constant k such that for each point $x \subset E_0[1] \cup e_0[1]$, either $g(x)$ or $g_0^2(x)$ is distance greater than k from $N_0[1] = N$. For λ sufficiently small, we have $d(g_0(x), g_\lambda(x)) < k/3$ for all x in N. Therefore $N_\lambda[3] \cap E_0[1] \cup e_0[1] = \emptyset$. Further, points in the set $E_\lambda[1] \cup e_\lambda[1]$ map by g_λ at least distance $2k/3$ outside N on the first or second iterate. Hence $N_0[3] \cap E_\lambda[1] \cup e_\lambda[1] = \emptyset$ and one can again use Proposition 3.B.1 to construct $\tilde{\pi}: N_\lambda^\#[1] \to N_0^\#[3]$. The statements that the diagrams almost commute can be derived from the convergence condition $g_\lambda(z) \to g_0(z)$ for all $z \in N$. ∎

The next theorem shows that maps $g_\lambda^\# \pi$ and $\pi g_0^\#$ are homotopic for λ sufficiently close to zero. It is necessary to assume that two sufficiently close maps having compact domains and ranges contained in N are homotopic. This is true if N is a compact smooth manifold with boundary because one can deform along geodesics. If N is a locally convex subset of some Euclidian space, then two maps a and b with ranges in N are homotopic via the homotopy $H(x, t) = ta(x) + (1 - t)b(x)$.

THEOREM 5.B.3: Suppose that the block N has the property that two sufficiently close maps having compact domains and ranges in N are homotopic. Then for λ sufficiently small, the diagrams in Theorem 5.B.2 commute up to homotopy.

Proof: Choose λ sufficiently small so that there exists a homotopy $H: (N_0[2] \cap N_\lambda[2]) \times [0, 1] \to N$ with $H_x(x, 0) = g_0(x)$ and $H(x, 1) = g_\lambda(x)$. Define a homotopy $G: N_0^\#[1] \times [0, 1] \to N_\lambda^\#[3]$ by the formula

$$G(x, t) = \begin{cases} \gamma H(x, t) \text{ if } x \in N_0[2] \cap N_\lambda[2] \\ * \text{ otherwise} \end{cases}$$

The function γ is the projection $\gamma: N_\lambda^\#[1] \to N_\lambda^\#[3]$. The function G is the desired homotopy. Similarly, one constructs a homotopy between the maps $g_0^\# \tilde{\pi}$ and $\tilde{\pi} g_\lambda^\#$. ∎

C. The Homology Conley Index

In section 5.A we assigned an index to a block. The goal here is to assign an index to an isolated invariant set by using a block which isolates the set. The index must be independent of the block used to isolate the invariant set, and the index must not change as the invariant set is continued. A nontrivial index must indicate that the associated invariant set is nonempty.

Definition: Let S^* be an isolated invariant set for a map f. Choose a nested sequence of isolating blocks $\{N_k\}$ with exit sets $\{E_k\}$ and exit threshold sets $\{e_k\}$ which converge to the set S^*. Require that N_k isolates S^* and also that $N_{k+1} \subset N_k - (E_k \cup e_k)$ for each k. Projection maps $\gamma_k: N_k^\# \to N_{k+1}^\#$ between index spaces are defined using Proposition 5.B.1 from section 5.A. The index map $f_k^\#$ of $N_k^\#$ induces homomorphisms for every $r \geq 0$ of the homology groups

THE CONLEY INDEX

$$(f_k^\#)_*: H_r(N_k^\#) \to H_r(N_k^\#)$$

also, for each k the index map $f_k^\#$ of $N_k^\#$ induces homomorphisms of the cohomology groups

$$(f_k^\#)^*: H^r(N_k^\#) \to H^r(N_k^\#)$$

These homomorhisms are called the *index homomorphisms on homology* and *cohomology*, respectively.

Definition: For a fixed nonnegative integer r consider the system of homology groups and homomorphisms $\{H_r(N_k^\#), (\gamma_k^j)_*\}$, where for $j > k$ the homomorphism $(\gamma_k^j)_*: H_r(N^\#[k]) \to H_r(N^\#[j])$ is defined by the equation $(\gamma_k^j)_* = (\gamma_{j-1} \ldots \gamma_k)_*$, and for $j = k$, $(\gamma_k^k)_*$ is the identity homomorphism. This system is a special case of what is called a "directed system". Let Σ denote the *direct sum* of the homology groups defined as follows: Let

$$M = \{\mu: Z^+ \to \cup \{H_r(N_k^\#): k \geq 0\}$$

The group Σ is the subset of M consisting of all functions μ which satisfy the conditions

$\mu(j) = 0$ except for finitely many j's and $\mu(j) \in H_r(N^\#[j])$ for each j.

A subgroup R of Σ is defined as

$$R = \left\{\mu \in \Sigma: \sum_{t=1}^{c} \gamma_t^c(\mu(t)) = 0, \text{ with } \mu(t) = 0 \text{ for } t > c\right\}$$

The *direct limit* DirectLimit$\{H_r(N_k^\#), (\gamma_k^j)_*\}$ of the directed system $\{H_r(N_k^\#), (\gamma_k^j)_*\}$ is defined to be the quotient group Σ/R.

Definition: The *homology Conley index in dimension* r of an isolated invariant set S^* is denoted by $CH_r(S^*)$. It is the direct limit of the homology groups of the directed sequence of homology groups and projection homomorphisms $\{H_r(N_k^\#), (\gamma)_*\}$. There is an induced index homomorphism $f^{\#\#}$ of $CH_r(S^*)$ which is defined using the homorphism of M induced by all the index maps $(f_k^\#)_*$. In particular, $f^{\#\#}(\mu)(t) = (f\#_t)_*(\mu(t))$.

Example: Let f be the linear map of the real line defined by $f(x) = 2x$. Then the interval $N = [-1, 1]$ is an isolating block for f. The space $N[j+1]$ is equal to the interval $[-2^{-j}, 2^{-j}]$. Each of the index space $N^\#[j]$ is topologically a circle with a distinguished base point. The homology group in dimension 1 or a circle is isomorphic to the group of integers Z. The projection map from $N^\#[j]$ to $N^\#[j+1]$ induces the identity map on homology groups. Thus the direct limit of the index groups can be identified with the direct limit of the sequence of groups $A_j = Z$, with homomorphisms $\gamma_j^{j+1} = id$. The direct sum Σ of the groups A_j can be thought of as the space of all vectors of infinite length having only finitely many nonzero integer entries, A natural "basis" for this sum is the sequence of vectors having 1 in the jth entry and zeros elsewhere. However, the difference between any two of the basis vectors is in the subgroup R. Therefore, the quotient group Σ/R is again isomorphic to the integers.

The previous example may seem like much ado about nothing. The next example is a little less simple.

Example: Let f be the logistic map of the real line defined by $f(x) = 9(x)(1-x)$. Let N be the closed interval $[-1, 2]$. Continuing the example from the previous section, we see that the index space $N^\#[j]$ is a "bouquet" of 2^j circles touching exactly at one point which is the base point. The homology group in dimension 1 of a bouquet of 2^j circles is the direct sum of 2^j copies of the integers which is denoted by A_j. In the direct sum Σ of these groups it is useful to look at those simple elements μ with $\mu(j) \neq 0$ for exactly one j. Further, given such an element, one may look at the least integer j for which there is a "least" element ν of the direct sum such that $\nu - \mu \in R$. The collection of the equivalence classes of these simple least elements generates the direct limit. For our example, the projection homomorphisms $\gamma_j^{j+1} : A_j \to A_{j+1}$ are injections. Hence each simple element μ determines a nonzero equivalence class in the direct limit Σ/R. The way this relates to the geometry is as follows: The set $N[j+1]$ consists of 2^j closed intervals, each of which corresponds to a circle in the bouquet of circles formed by the index space $N^\#[j]$. The kth circle "carries" a homology class $z(j, k)$ such that the collection of these classes $\{z(j, k): 1 \leq k \leq 2^j\}$ forms a basis for the group A_j. Each simple element $\mu[j, k] \in \Sigma$ with $\mu[j, k](j) = z(j, k)$ determines a distinct nonzero element in the direct limit. Thus every interval in the set $N[j+1]$ determines a nonzero element in the direct limit, and this reflects the complexity of the invariant Cantor set isolated by N.

THEOREM 5.C.1: The homology Conley index of an isolated invariant set S^* is independent of the choice of the sequence of isolating blocks used to define the index.

Proof: Let $\{N_k\}$ and $\{\hat{N}_k\}$ be any two sequences of isolating blocks which converge to S^*. For each integer k there is an integer $t > k$ such that $N_t \subset \hat{N}_k - (\hat{E}_k \cup \hat{e}_k)$. Thus, using Proposition 5.B.7, there is a projection map $\hat{\Gamma}_k : \hat{N}_k^\# \to N_t^\#$. Similarly, there is a projection map $\Gamma_k : N_k^\# \to \hat{N}_s$ for some $s > k$. These projection maps commute with the projection maps

$\gamma\colon N_k^\# \to N\#_m$ and $\hat{\gamma}_k^m\colon \hat{N}_k^\# \to \hat{N}_m^\#$. At the homology level, the induced projection homomorphisms between index spaces also commute. Consequently, the direct limits are isomorhpic. The isomorphism β is constructed as follows: Choose a homology class $z \in H_r(N_k^\#)$. This cycle determines an equivalence class $[z]$ in the direct limit $CH_r(S^*)$. Generate a new homology class w with $w = (\Gamma_k)_*(z) \in H_r(\hat{N}_s^\#)$. Then the equivalence class of w in $CH_r(S^*)$ is defined to be $\beta([z])$. One must check that the function β is well defined and is an isomorphism. This is left as an exercise. ∎

The index maps also commute with the various projections, and there are homotopies of the right sort via Theorem 5.B.2, so that the induced index homomorphism is also independent of the sequence of blocks. However, since the definition of the Conley index involves a sequence of blocks, it is not generally computable. The next result involves just one block and its index homomorphism.

Definition: An element x of a group G is *persistent* relative to a homomorphism h of G if $h^n(x)$ is not equal to the identity element of the group for each positive integer n.

THEOREM 5.C.2: Let N be an isolating block for a map f. If α is a persistent element of the index group $H_r(N^\#)$ relative to the index homomorphism, then the set of orbits trapped by N is nonempty.

Proof: From the commutativity of the diagram in Figure 5.1 and from Theorem 5.A.2 we have $(f^\#)^3 = \phi_2\phi_3\phi_4\gamma_3\gamma_2\gamma_1$. There is a similar diagram with the index spaces replacd by homology groups in dimension r, and the maps replaced by the associated homomorphisms between homology groups. Since α is a persistent cycle, $(f^\#)_*^3(\alpha)$ is not zero and therefore $(\gamma_3\gamma_2\gamma_1)_*(\alpha)$ is not zero. Hence the homology group $H_r(N^\#[4])$ is nontrivial. Hence $N[4]$ is not the empty set. This argument generalizes to show that $N[k]$ is nonempty for each k. Therefore $N[\infty]$ is nonempty. ∎

THEOREM 5.C.3: Let f_λ be a one-parameter family of maps for $0 \le \lambda \le b$, and suppose that N is an isolating block for each map in this family. Let $\alpha \in H_r(N_0^\#[1])$ be a persistent cycle. Then for λ sufficiently small, the cycle α has a local continuation $\beta \in H_r(N_\lambda^\#[3])$, and β is a persistent cycle with respect to the index homomorphism $(f_\lambda^\#)_*$.

Proof: The cycle $(f_0^\#)_*^k(\alpha)$ is also a persistent cycle and consequently, by Theorem 5.C.2, $(\gamma_1^5)_*(f_0^\#)_*^k(\alpha) \ne 0$. Using an argument similar to the one in Theorem 5.C.2, for λ sufficiently small, there are a projection maps $\pi\colon N_0^\#[1] \to N_\lambda^\#[3]$ and $\rho\colon N_\lambda^\#[3] \to N_0^\#[5]$ as pictured in Figure 5.2. Let $\beta = (\pi)_*(\alpha)$. To show that β is a persistent cycle it is sufficient to show that $(f_\lambda^\#)_*^k(\beta) \ne 0$ for each positive integer k. Since the diagram of maps in Figure

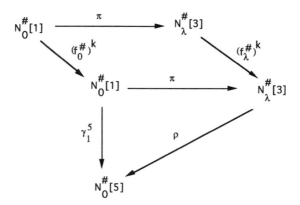

Figure 5.2. Commutative diagram of index spaces and maps for the proof of Theorem 5.C.3.

5.2 is commutative, we have $(\rho)_*(f_\lambda^\#)_*^k(\beta) = (\gamma_1^5)_*(f_0^\#)_*^k(\alpha)$, and therefore $(f_\lambda^\#)_*^k(\beta) \neq 0$. ∎

Note that different cycles may continue over different parameter ranges, thus reflecting internal changes in the structure of the continued invariant set.

Remark: Attractor blocks have not been assigned an index since their exit sets are empty. It is natural to assign the attractor block itself as its own index space. For an attractor block N one may study the homomorphisms on homology groups $f^*: H_r(f^j(N) \to H_r(f^{j+1}(N))$ and again construct direct limits. A good example to keep in mind in this context is the solenoid example from section A of Chapter 4.

D. References

Isolating blocks and their homology groups were used to study invariant sets for flows in the fundamental paper of Conley and Easton (1971). Conley first defined an index for isolated invariant sets for flows in Conley (1978). An index of isolated invariant ses for maps was developed in Robin and Salamon (1988), using "index pairs" and "shape theory." Isolating blocks for maps were introduced in Easton (1989). Mrozek defined the Conley index for invertible maps in Mrozek (1990) and Mrozek and Rybakowski (1991). The approach to the Conley index presented here is new. The Conley memorial volume (Herman et al. 1988) has a brief history of the development of the Conley index and contains several chapters which contribute to the theory of this index. See also Mischaikow (1995).

6

Symplectic Maps

Symplectic maps are the discrete analog of Hamiltonian dynamical systems. They are used to model dynamics when there is no friction or dissipation of energy. The fixed-time map of a Hamiltonian flow is a symplectic map. Symplectic maps are volume preserving, and preserve a symplectic structure. Symplectic structures are defined in terms of differential forms. The reader will find a brief discussion of tensors, differential forms, and manifolds in Appendix C. Symplectic manifolds are discussed in Appendix D.

A. Linear Symplectic Maps

Definition: A *symplectic vector space* is a vector space V (of even dimension) together with a function $\omega: V \times V \to R^1$. The function ω must be linear in each entry, and must have the property that $\omega(v, w) = -\omega(w, v)$. The function ω is further required to be nondegenerate: For each nonzero vector v, there exists a vector w such that $\omega(v, w) = 1$. Any function with these properties is called a *nondegenerate exterior 2-form*.

A *symplectic basis* for V is a basis $\{b^1, \ldots, b^{2n}\}$ for V such that

$$\omega(b^j, b^{j+n}) = 1 \quad \text{for } 1 \leq j \leq n \quad \text{and} \quad \omega(b^j, b^k) = 0 \quad \text{if } k \neq j+n$$

Definition: V^{2n} is the vector space of real column vectors of height $2n$.

Example: The *standard symplectic structure* on V^{2n} is given by the exterior 2-form

$$\omega: V^{2n} \times V^{2n} \to R^1; \quad \Omega(v, w) = v^t J w$$

The row vector v^t denotes the transpose of the column vector v, and the symbol I represents the $n \times n$ identity matrix, and J is the $2n \times 2n$ matrix

$$J = \begin{pmatrix} 0 & I \\ -I & 0 \end{pmatrix}$$

The standard basis for V^{2n} is a symplectic basis with respect to the standard 2-form Ω.

Definition: If S is a subspace of a symplectic vector space (V, ω) then the subspace of V which is *skew orthogonal* to S is

$$S^\perp = \{v: \omega(s, v) = 0 \text{ for all } s \in S\}.$$

THEOREM 6.A.1: Each $2n$-dimensional symplectic vector space (V, ω) has a symplectic basis.

Proof: Choose any nonzero vector b^1 for the first basis vector. Since ω is nondegenerate, there must be a vector b^{n+1} such that $\omega(b^1, b^{n+1}) = 1$. Since the 2-form ω is anti- or skew-symmetric, it follows that b^1 and b^{n+1} are linearly independent. Let S be the two-dimensional subspace of V spanned by b^1 and b^{n+1}. Let $S^\perp$ be the subspace of V which is skew orthogonal to S. Define a linear transformation

$$L: V \to R^2; v \to (\omega(b^1, v), \omega(b^{n+1}, v))$$

The transformation L is onto, and the kernel of L is $S^\perp$. A theorem from linear algebra states that the dimension of the kernel of a linear transformation plus the dimension of its range is equal to the dimension of the (finite-dimensional) vector space which is its domain. Therefore $\dim(S^\perp) + 2 = \dim(V)$. Consider the restriction of ω to $S^\perp$. Since ω is nondegenerate, it follows from the definition of $S^\perp$ that the restriction of ω to $S^\perp$ is also nondegenerate. One can now construct the next pair of symplectic basis vectors as a pair of vectors in $S^\perp$. Continue this construction until a symplectic basis for V is constructed. ∎

Definition: A *linear symplectic map* of a symplectic vector space (V, ω) is a linear transformation L such that for all vectors v, w, $\omega(Lv, Lw) = \omega(v, w)$.

Definition: A $2n \times 2n$ matrix A is a *symplectic matrix* if $\Omega(Av, Aw) = \Omega(v, w)$ for all vectors v, w. The set of all such matrices forms a group called the *symplectic group* which is denoted by $Sp(2n)$. Each matrix in $Sp(2n)$ defines a linear symplectic map of (V^{2n}, Ω). To understand such maps it helps to find their eigenvalues and eigenvectors.

THEOREM 6.A.2: Any symplectic vector space V of dimension $2n$ is isomorphic to (V^{2n}, Ω) with a symplectic isomorphism.

Proof: Define the isomorphism to take a symplectic basis for V which exists by Theorem 6.A.1 to the standard basis for V^{2n}. ∎

SYMPLECTIC MAPS

PROPOSITION 6.A.3: A matrix M represents a symplectic linear transformation of V^{2n} (or of C^{2n}) if and only if $M^t JM = J$.

Proof: M is symplectic if and only if $\Omega(Mv, Mw) = \Omega(v, w)$. From the definition of Ω, the matrix M is symplectic if and only if $\bar{v}^t M^t JMw = \bar{v}^t Jw$ for all possible vectors v and w. Therefore, M is symplectic if and only if $M^t JM = J$. ∎

Remark: The equation $M^t JM = J$ is equivalent to the equation $-JM^t J = M^{-1}$ since $J^{-1} = J^t = -J$. This makes the inverse of M easy to compute!

PROPOSITION 6.A.4: Suppose M is a linear symplectic map of V^{2n} and suppose R and S are the generalized eigenspaces $R = \{v : (M - \lambda I)^{2n} v = 0\}$ and $S = \{v : (M - \mu I)^{2n} v = 0\}$. If $\lambda\mu \neq 1$, then R and S are skew-orthogonal.

Proof: The proof is by induction. Assume that the equations $(M - \lambda I)^a v = 0$, and $(M - \mu I)^b w = 0$ imply that $\Omega(v, w) = 0$ whenever $a + b < k$. This is true for $k = 1$. It is sufficient to show that $\Omega(v, w) = 0$ when $a + b = k$. Let $(M - \lambda I)^{a-1} v = v^*$, and $(M - \mu I)^{b-1} w = w^*$. Then $Mv - v^* = \lambda v$, and $Mw - w^* = \mu w$. Therefore,

$$\lambda\mu\Omega(v, w) = \Omega(Mv - v^*, Mw - w^*)$$
$$= \Omega(Mv, Mw) + \Omega(-v^*, Mw) + \Omega(Mv, w^*) + \Omega(v^*, w^*)$$

Since M is symplectic, $\Omega(Mv, Mw) = \Omega(v, w)$. Since $(M - \lambda I)^a Mv = M(M - \lambda I)^a v = 0$, it follows from the induction assumption that $\Omega(Mv, w^*) = 0$. Similarly, $\Omega(-v^*, Mw) = 0$ and $\Omega(v^*, w^*) = 0$. Therefore, $(\lambda\mu - 1)\Omega(v, w) = 0$. ∎

Definition: A subspace W of V^{2n} is a *symplectic subspace* if the 2-form Ω restricted to W is nondegenerate.

PROPOSITION 6.A.5: Suppose M is a symplectic transformation of V^{2n} and suppose R and S are the generalized eigenspaces $R = \{v : (M - \lambda I)^{2n} = 0\}$ and $S = \{v : (M - \mu I)^{2n} = 0\}$. If $\lambda\mu = 1$, then the subspace T spanned by both R and S is a symplectic subspace.

Proof: It is sufficient to show that for a fixed v in T, $\Omega(v, w) = 0$ for all w in T implies that $v = 0$. Let L be the subspace spanned by all the generalized eigenvectors with eigenvalues not equal to λ or μ. If $\Omega(v, w) = 0$ for all w in T, then, by Proposition 6.A.4, $\Omega(v, w) = 0$ for all u in L. Since the generalized eigenspaces span V^{2n}, $\Omega(v, w) = 0$ for all vectors z in V^{2n}. Since Ω is nondegenerate, $v = 0$. ∎

PROPOSITION 6.A.6: Suppose M is a symplectic transformation of V^{2n}. If $p(\lambda)$ is the characteristic polynomial of M, then $p(\lambda) = \lambda^{2n}p(\lambda^{-1})$. Hence, the eigenvalues of M come in quartets $\lambda, \lambda^{-1}, \bar{\lambda}, \bar{\lambda}^{-1}$.

Proof

$$p(\lambda) = \det(M - \lambda I) = \det(M^t - \lambda I) = \det(-JM^{-1}J + \lambda JJ)$$
$$= \det(-JM^{-1}J + \lambda JJ) = \det J \det(-M^{-1} + \lambda I) \det J = \det(-M^{-1} + \lambda I)$$
$$= \det(-M^{-1} + \lambda I) = \det(M^{-1})\det(-I + \lambda M)$$
$$= \det(\lambda[M - \lambda^{-1}I]) = \lambda^{2n}p(\lambda^{-1})$$ ∎

THEOREM 6.A.7: If M is a symplectic matrix, then the determinant of M is equal to 1.

Proof: Recall that the determinant of a product of matrices is the product of their determinants. Therefore, it follows from the equation $M^t J M = J$ that $\det(M) = \pm 1$. To show that M has determinant equal to 1 is more difficult. One approach is to use the formalism of differential forms discussed in Appendix C. Since M preserves the standard 2-form Ω, it also preserves the n-fold wedge Ω^n of this form with itself. Therefore, $\Omega^n(Me_1, \ldots, Me_{2n}) = \Omega^n(e_1, \ldots, e_{2n}) = 1$, where $\{e_1, \ldots, e_{2n}\}$ is the standard basis of V^{2n}. It follows from the definitions of Ω and Ω^n that $\Omega^n(Me_1, \ldots, Me_{2n}) = \det M$. ∎

THEOREM 6.A.8: Suppose M is a symplectic transformation of V^{2n}. If the eigenvalues of M are distinct, then there is a complex symplectic matrix A such that

$$A^{-1}MA = \text{diag}(\lambda_1, \ldots, \lambda_n, \lambda_1^{-1}, \ldots, \lambda_n^{-1})$$

Proof: Choose $v_j \neq 0$ so that $Mv_j = \lambda_j v_j$ for $1 \leq j \leq n$ and choose $w_j \neq 0$ so that $Mw_j = \lambda_j^{-1} w_j$ for $1 \leq j \leq n$. The subspace spanned by v_j and w_j is symplectic by Proposition 6.A.5. Replace w_j by a scalar multiple if necessary so that $\Omega(v_j, w_j) = 1$. Put these vectors together to form the columns of the matrix $A = (v_1, \ldots, v_n, w_1, \ldots, w_n)$. A is symplectic because the v's and w's form a symplectic basis for V^{2n}. $A^{-1}MA$ is diagonal because the v's and w's are eigenvectors. ∎

Definitions: A subspace S of a symplectic vector space (V, ω) is *isotropic* if $S \subset S^{\perp}$. The subspace S is *coisotropic* if $S^{\perp} \subset S$, and the subspace is *Lagrangian* if $S^{\perp} = S$.

The technical terms "isotropic" and "coisotropic" are defined for completeness. Lagrangian subspaces and "Lagrangian manifolds" will play an important role in subsequent sections.

B. Classical Mechanics

Classical mechanics is based on Newton's law that force equals mass times acceleration. In modeling a general solar system, Newton studied a system of n particles moving in three-dimensional space. The state of the system consists of all the positions and velocities of the particles. It is remarkable that the state does not depend on the accelerations or on higher derivatives of the particle motions. This makes Newton's law a statement about how the world works. Newton's law is written as a second-order system of ordinary differential equations:

$$m_k \ddot{x}_k = F_k(x, \dot{x}, t); \; x_k \in R^3, \qquad x = (x_1, \ldots, x_n), \, 1 \leq k \leq n$$

These equations can be rewritten using vector notation as

$$\dot{x} = v, \qquad M\dot{v} = F(x, v, t)$$

where M is a diagonal matrix with the masses down the diagonal.

A *holonomic force* is a force which does not depend on velocities, and has the property that at each fixed time, the work done in moving a test particle around a closed loop in x-space (configuration space) is zero. Equivalently, the force is holonomic if $F(x, t) = \text{grad } U(x, t)$ for some real-valued function $U(x, t)$. The function U is called the *potential energy* of the system. If the force is holonomic, then these equations can be written as a first-order system:

$$\dot{x} = M^{-1} y, \qquad \dot{y} = -\text{grad } U(x, t)$$

where $y = Mv$. One defines a function $H(x, y) = \frac{1}{2} y^t M y + U(x, t)$, which represents the total energy of the system, and the first-order system of equations now has the form

$$\dot{x}_j = \frac{\partial H(x, y, t)}{\partial y_j}, \qquad \dot{y}_j = \frac{-\partial H(x, y, t)}{\partial y_j}$$

Any system of differential equations having this form is called a *Hamiltonian system* and the function H is called the *Hamiltonian* of the system.

THEOREM 6.B.1: If the Hamiltonian function H does not depend on time, then total energy $H(x, y)$ is conserved.

Proof: Suppose that $(x(t), y(t))$ is a solution of the Hamiltonian system. To show that $H(x(t), y(t))$ is constant in time t, it is sufficient to show that the time derivative is zero. By the chain rule,

$$\frac{d}{dt} H(x(t), y(t)) = \sum_j \frac{\partial H(x(t), y(t))}{\partial x_j} \dot{x}_j + \frac{\partial H(x(t), y(t))}{\partial y_j} \dot{y}_j$$

$$= \sum_j \frac{\partial H(x(t), y(t))}{\partial x_j} \frac{\partial H(x(t), y(t))}{\partial y_j}$$

$$- \frac{\partial H(x(t), y(t))}{\partial y_j} \frac{\partial H(x(t), y(t))}{\partial x_j} = 0 \quad \blacksquare$$

Liouville's theorem (Theorem 6.B.4) states that the general solution or flow of a Hamiltonian system of differential equations is a one-parameter family of symplectic maps. The proof of this theorem requires the following definition and two lemmas.

Definition: A *symplectic map* or R^{2n} is a map whose Jacobian matrix at each point is a symplectic matrix.

Remark: Since a symplectic matrix has determinant 1, it follows that any symplectic map preserves volume.

LEMMA 6.B.2: Let $\psi(t)$ denote the fundamental matrix solution to the linear system of differential equations $\dot{z} = JS(t)z$ where $S(t)$ is a $2n \times 2n$ symmetric matrix which depends continuously on t. If $\psi(0)$ is the identity matrix, then $\psi(t)$ is a symplectic matrix for all t.

Proof: Since $\psi(t)$ is a fundamental matrix, it satisfies the matrix differential equation $\dot{\psi} = JS\psi$. Because $\psi(0)$ is the identity matrix, we have $[\psi(0)]^t J \psi(0) = J$. Hence, to show that $[\psi(t)]^t J \psi(t) = J$ it is sufficient to differentiate the left side of this equation with respect to t and show that the result is zero. Recall that the product rule for differentiation works for matrices if one keeps the matrices in the same order. We calculate

$$\frac{d}{dt}\{[\psi]^t J \psi\} = [\dot{\psi}]^t J \psi + [\psi]^t J \dot{\psi} = [JS\psi]^t J \psi + [\psi]^t J(JS\psi)$$

$$= [\psi]^t S^t J^t J \psi + [\psi]^t JJS\psi$$

Since S is symmetric, and $J^t J$ is the identity matrix, and JJ is minus the identity matrix, the result is zero. $\blacksquare$

A common practice in the study of differential equations is to "linearize" the equations along a reference orbit. We use this procedure to show that the flow of a Hamiltonian system of differential equations is a one-parameter group of symplectic maps.

Notation: Let $G: R^k \times R^m \times R^l \to R^n$ be a smooth function. The symbol $D_1 G(x, y, z)$ denotes the matrix of partial derivatives whose i, jth entry is $\partial/\partial x_i \{G_j(x, y, z)\}$. The symbol $D_2 G(x, y, z)$ denotes the matrix of partial derivatives whose i, jth entry is $\partial/\partial y_i \{G_j(x, y, z)\}$.

LEMMA 6.B.3: Consider a smooth system of differential equations $\dot{z} = F(z, t), \dot{t} = 1$. Suppose that $\Phi: R^m \times R^l \times R^l \to R^m \times R^l$; $\Phi(z, s, t) = (\phi(z, s, t), s + t)$ is the smooth flow determined by this system. Then the family of Jacobian matrices $\psi(z, t) = D_i \phi(z, 0, t)$ solves the matrix system of differential equations $\dot{\psi} = D_1 F(\phi(z, 0, t), t) \psi$ with initial condition $\psi(0) = I$.

Proof: We start by calculating the rate of change of the i, jth entry in the matrix Ψ.

$$\frac{d}{dt} \psi_{i,j}(z, t) = \frac{d}{dt} \left\{ \frac{\partial}{\partial z_j} \phi_i(z, 0, t) \right\} = \frac{\partial}{\partial z_j} \left\{ \frac{d}{dt} \phi_i(z, 0, t) \right\}$$

By hypothesis $d/dt\, \phi_i(z, 0, t) = F_i(\phi(z, 0, t), t)$. Hence,

$$\frac{d}{dt} \psi_{i,j}(z, t) = \sum_{k=1}^{m} \frac{\partial}{\partial z_k} F_i(\phi(z, 0, t), t) \left[\frac{\partial}{\partial z_j} \phi_k(z, 0, t) \right]$$

The sum on the right is exactly the i, jth entry in the matrix $D_1 F(\phi(z, 0, t), t) \psi(z, t)$. Therefore, $\psi(z, t)$ solves the matrix differential equation. The initial condition $\psi(0) = I$ is satisfied because $\phi(z, 0, 0) = z$. ∎

THEOREM 6.B.4 (Liouville's Theorem): Suppose that $\Phi: R^{2n} \times R^l \times R^l \to R^{2n} \times R^l$ is a differential flow generated by the system of differential equations

$$\dot{z} = JD_z H(z, t), \quad \dot{t} = 1$$

We write $\Phi(z, t) = (\phi(z, t)), t)$. Then, for fixed t, the map $f(z) \equiv \phi(z, t)$ is symplectic.

Proof: The Jacobian matrix $D\phi(z, t)$ can be calculated using Lemma 6.B.3. It solves the matrix differential equation $\dot{\psi} = JS(t)\psi$ with $S_{i,j}(t) = \partial/\partial z_i \{\partial/\partial z_j H(\phi(z, t))\}$. By Lemma 6.B.2, the Jacobian matrix $D\phi(z, t)$ is a symplectic matrix. Therefore, the map f is symplectic. ∎

C. Variational Principles

Variational principles form the basis for most of the fundamental physical laws.

In continuous-time dynamics, an example that appeals to our intuition is the geometric optics model for light. In this model a photon travels in such a way as to minimize the time to go between any two points on its path. The velocity of light depends on the index of refraction of the material that it passes through. Thus one can derive Snell's law for the path of a beam of light traveling in air and then through a flat pane of glass.

Geodesics on a surface provide another example. Suppose that M is a smooth surface in space. A smooth curve contained in M is a *geodesic* provided that, given any two points on the curve, the given curve has the shortest arc-length among all nearby curves which join the two points.

One may view the Lagrangian formulation of mechanics as giving a way to select orbits by means of a variational principle. An orbit must "locally" be an extremal of a given "action function." One can show that extremals must be solutions of a second-order system of ordinary differential equations called the *Euler–Lagrange equations*. Using the notation of the previous section (see p.109), these equations are written

$$\frac{d}{dt}[D_2 L(q, \dot{q})] = D_1 L(q, \dot{q}) \qquad \text{for } 1 \leq k \leq n$$

In Lagrangian mechanics, the Lagrange function L depends on position and velocity. To formulate analogous equations when time is discrete, we replace velocity by average velocity. Then we have a Lagrange function which depends on the initial position a and average velocity $b - a$, where b is the "new position" after the passage of one time unit. Thus we assume that

$$L: R^n \times R^n \to R^1$$

is a given Lagrange function. A *discrete path* in R^n from a to c is a function

$$\gamma: [t_0, t_1] \to R^n$$

where $[t_0, t_1]$ is an interval in the set of integers Z, $\gamma(t_0) = a$ and $\gamma(t_1) = b$. Let $P = P(a, c, t_0, t_1)$ denote the set of all discrete paths joining a to c. Suppose that the *length* $t_1 - t_0$ of any path is at least 3.

Definitions: The *action function* $A: P \to R^1$ is defined by the formula

$$A(\gamma) = \sum L(\gamma(j), \gamma(j+1)) \qquad \text{where } t_0 \leq j < t_1$$

We view the action function as a function on R^{dn} where $d = t_1 - t_0 - 1$.

A path γ is *an extremal of the action function* if the point $u = (\gamma(t_0 + 1), \ldots, \gamma(t_1 - 1))$ is a critical point of A. This means that all partial derivatives of A vanish at u. The *discrete Euler–Lagrange equations* (or variational equations) are therefore

$$D_1 L(\gamma(j), \gamma(j+1)) + D_2 L(\gamma(j-1), \gamma(j)) = 0 \quad \text{for } t_0 < j < t_1$$

We can view these equations as defining a subset Φ of $R^n \times R^n \times R^n \times R^n$.

$$\Phi = \{(a, b, b, c) : D_2 L(a, b) + D_1 L(b, c) = 0\}$$

If the set Φ is the graph of a map $(a, b) \to (b, c)$ from $R^n \times R^n$ to $R^n \times R^n$, then the extremal path is obtained by iterating this map. This is the discrete analog of solving the Euler–Lagrange system of differential equations.

In passing from the Lagrangian to the Hamiltonian formulation of classical mechanics, the Legendre transformation is used. We also use a Legendre transformation to introduce a new "momentum" variable. Then we demonstrate how symplectic maps arise from the transformed discrete Euler–Lagrange equations.

Definition: The *Legendre transformation* associated with L is a map

$$g : R^n \times R^n \to R^n \times R^n; (a, b) \to (b, D_2 L(a, b)).$$

Suppose instead one transforms $R^n \times R^n \times R^n \times R^n$ by $g \times g$. This means that Φ transforms to the set

$$\Lambda = \{(b, -D_1 L(b, c)), (c, D_2 L(b, c))]$$

The set Λ can be equivalently represented by an equation involving 1-forms. Consider the 1-form

$$\sum_{j=1}^{n} P_j \, dQ_j - p_j \, dq_j$$

on $R^n \times R^n \times R^n \times R^n$ with coordinates (q, p, Q, P). This 1-form will subsequently be denoted by $P \, dQ - p \, dq$. Define a function S on $R^n \times R^n \times R^n \times R^n$ by $S(q, p, Q, P) = L(q, Q)$. Then

$$\Lambda = \{(q, p, Q, P) : P \, dQ - p \, dq = dS\}$$

In terms of the differential form expression, a remarkable property of the set Λ is apparent.

PROPOSITION 6.C.1: The integral of $P\,dQ - p\,dq$ around a closed loop Γ which is contained in the set Λ is zero.

Proof: The integral of dS around any closed loop is zero. ∎

If the set L is the graph of a map, then the following two propositions show that the map is symplectic.

PROPOSITION 6.C.2: Suppose that Λ is the graph of a function f. Then there is a function G such that $f^*p\,dq - p\,dq = dG(q,p)$.

Proof: Choose a closed loop γ in $R^n \times R^n$ and let $\Gamma(t) = (\gamma(t), f(\gamma(t)))$. Then

$$\int_\lambda f^*p\,dq - p\,dq = \int_\Gamma P\,dQ - p\,dq = \int_\Gamma dS = 0$$

Thus one can define

$$G(a, b) = \int_\lambda f^*p\,dq - p\,dq$$

where $\lambda(t) = (ta, tb)$ for $0 \leq t \leq 1$ is the straight path joining the origin to (a, b). ∎

Definition: A map f of R^{2n} is said to be an *exact symplectic* map if there exists a function G such that $f^*p\,dq - p\,dq = dG(q,p)$.

PROPOSITION 6.C.1: Any exact symplectic map is a symplectic map.

Proof: Exterior differentiation commutes with pull-back. Thus

$$0 = d\,d(G) = d(f^*\mu - \mu) = f^*\,d\mu - d\mu = f^*\Omega - \Omega.$$ ∎

Example: The motion of a simple pendulum is modeled by the Lagrange function $L(q, \dot{q}) = \frac{1}{2}(\dot{q})^2 - (1 - \cos(q))$. A discrete version of this function is $L(q, Q) = \frac{1}{2}(Q - q)^2 - (1 - \cos(q))$, where q stands for the current position and Q for the future position of the pendulum. One can view the set $\Lambda = \{(q, -D_1L(q, Q)), (Q, D_2L(q, Q))\}$ as the graph of a transformation $(q, p) \to (Q, P)$. Thus

$$p = -D_1L(q, Q) = Q - q + \sin(q) \quad \text{and} \quad P = D_2L(q, Q) = Q - q$$

Solving for Q and P in terms of p and q, we have a version of the standard map:

$$Q = q + p - \sin(q), \qquad P = p - \sin(q)$$

D. Generating Functions

We have seen that symplectic maps arise from discrete variational principles and we have seen that the flow generated by an autonomous Hamiltonian system of differential equations is a one-parameter group of symplectic maps. The goal of this section is to show how to produce symplectic maps. One obvious method is to write down Hamiltonian functions and solve the associated sets of differential equations. Except for rather special Hamiltonian functions, this is not possible since it involves explicitly solving a nonlinear system of differential equations. The classic way to produce symplectic maps is to determine the map implicitly by using a generating function.

For example, in Proposition 6.C.3 the mapping arising from a discrete Lagrange variational principle was shown to be exact symplectic. The generating function G in this case is related to the Lagrangian function L by $G(q, p) = L(g^{-1}(q, p))$, where g is the Legendre transformation.

From a modern point of view, the graphs of symplectic maps are "Lagrangian submanifolds." In order to generate symplectic maps, we first show how to produce Lagrangian submanifolds.

Definition: The graph L of a smooth function $\Lambda : R^n \to R^n$ is a *Lagrangian submanifold* of R^{2n} if the 2-form $dq \wedge dp$ vanishes on L. Thus $\Omega(u, v) = 0$ whenever u and v are vectors tangent to L. Alternately, each tangent plane to L must be a Langrangian subspace of (V^{2n}, Ω).

THEOREM 6.D.1: Suppose that the graph L of a smooth function $\Lambda: R^n \to R^n$ is a Lagrangian submanifold of R^{2n}. Then there exists a smooth function $S: R^n \to R^1$ such that

$$\Lambda(x) = \left(\frac{\partial S(x)}{\partial x_1}, \ldots, \frac{\partial S(x)}{\partial x_n}\right)$$

Proof: Define the function $S(x)$ by the formula

$$S(x) = \int_\gamma p \, dq$$

where γ is the curve $\gamma(t) = \Lambda(tx); 0 \leq t \leq 1$. The curve γ is contained in L. Since L is Lagrangian, by Stokes's theorem the integral of the 1-form $p \, dq$ along any curve from $\Lambda(0)$ to $\Lambda(x)$ is equal to $S(x)$. To show that

$$\Lambda_j(x) = \frac{\partial S(x)}{\partial x_j}$$

choose a smooth curve $u(t)$ from 0 to x in R^n with $u(0) = 0, u(1) = x$, and $u(t) = x + te_j$ for t in a neighborhood of 1. Then

$$\frac{\partial S(x)}{\partial x_j} = \frac{\partial}{\partial t}\int_0^t \Lambda_j(u(t))\,dt|_{t=1} = \Lambda_j(u(1)) = \Lambda_j(x) \quad \blacksquare$$

Note that in Theorem 6.D.1 the Lagrangian submanifold L is given by the equation $L = \{p\,dq - dS = 0\}$. This motivates the following definition (which is a generalization of the standard definition).

Definition: A function G on R^{2n} is a *generating function* for a Lagrangian submanifold L if there exists a 1-form α such that

$$d\alpha = \sum_{k=1}^{n} dp_k \wedge dq_k$$

and the set L is given by the equation $\alpha - dG = 0$.

If L is a Lagrangian submanifold of a symplectic manifold (M, ω), then the tangent plane to L at a point m is a Lagrangian subspace of the tangent plane to M at that point. By Darboux's theorem, one can choose symplectic coordinates (q, p) in a neighborhood of m such that L is tangent to the set $\{p = 0\}$ at the point $(0, 0)$. Thus L is the graph of a function $\Lambda(q)$. By Theorem 6.D.1, Λ is locally defined by a generating function S.

The following theorem shows that submanifolds defined by a generating function are Lagrangian submanifolds.

THEOREM 6.D.2: Suppose that G is a smooth function $G: R^{2n} \to R^1$. Let α be a 1-form on R^{2n} such that $d\alpha = \Omega$. If the set $L = \{(q, p) : dG(q, p) - \alpha(q, p) = 0\}$ is a submanifold of R^{2n}, then it is a Lagrangian submanifold of R^{2n}.

Proof: It is sufficient to show that the integral of Ω over any smooth disk embedded in L is zero. By Stokes's theorem it is sufficient to show that the integral of α around any closed loop contained in L is zero. Since $d(G) = \alpha$ on L, it is sufficient to show that the integral of dG around any loop is zero. This is true since by Stokes's theorem such an integral is equal to the difference of the function G evaluated at the endpoint and the starting point of the loop. $\blacksquare$

THEOREM 6.D.3: Suppose that g is a symplectic map of R^{2n}. Suppose that $L = \{(q, p): dG(q, p) - \alpha(q, p) = 0\}$ is a Lagrangian submanifold with generating function G. Then $g^{-1}(L)$ is a Lagrangian submanifold given by

$$g^{-1}(L) = \{(q,p): dG(g(q,p)) - g^*\alpha(q,p) = 0\}$$

Proof: Since exterior differentiation commutes with the pull-back operation, $dG(g(q,p)) = g^* dG(q,p) = g^*\alpha(q,p)$. Further, $dg^*\alpha = g^* d\alpha = g^*\Omega = \Omega$ and therefore the composite function $G(g(q,p))$ is a generating function with respect to the 1-form $g^*\alpha$. ∎

Definition: Given the symplectic space $(R^{2n}, dq \wedge dp)$, the Cartesian product manifold $R^{2n} \times R^{2n}$ with coordinates (q, p, Q, P) has a *product symplectic structure* $dQ \wedge dP - dq \wedge dp$.

THEOREM 6.D.4: A smooth map f of R^{2n} is symplectic if and only if the graph of f is a Lagrangian submanifold of $(R^{2n} \times R^{2n}, dQ \wedge dP - dq \wedge dp)$.

Proof: The statement that f is a symplectic map is equivalent to the equation $f^*(dq \wedge dp) = dq \wedge dp$. This equation is in turn equivalent (by Stokes's theorem) to the condition that for each smooth path $\gamma: [0,1] \to R^{2n}$ with $\gamma(0) = \gamma(1)$ the equation

$$\oint_\mu p\,dq = \oint_\gamma p\,dq$$

is satisfied where $(\mu(t) = f(\gamma(t))$ for $t \in [0,1]$. Now form the loop Γ in R^{4n} defined by $\Gamma(t) = (\gamma(t), f(\gamma(t)))$. It follows from the preceeding equation that

$$\oint_\Gamma P\,dQ - p\,dq = 0$$

By Stokes's theorem, the statement that this equation holds (for all loops Γ in the graph of f) is equivalent to the statement that the 2-form $dQ \wedge dP - dq \wedge dp$ vanishes on the graph of f. ∎

Definition: A function G on $R^n \times R^n \times R^n \times R^n$ is a *generating function* for a symplectic map f of R^{2n} if there exists a 1-form α such that $d\alpha = dP \wedge dQ - dp \wedge dq$, and the graph of f is given by the equation $\alpha - dG = 0$.

One can use this definition to get different types of generating functions. There are four classical choices of α:

$$\alpha_1 = Q\,dP - q\,dp, \qquad \alpha_2 = Q\,dP + p\,dq,$$
$$\alpha_3 = -P\,dQ - q\,dq, \qquad \alpha_4 = -P\,dQ + p\,dq$$

leading to four types of generating functions $G_1(p, P)$, $G_2(q, P)$, $G_3(p, Q)$, $G_4(q, Q)$.

To show that every symplectic map near the identity map has a generating function we use Theorems 6.D.1 and 6.D.4.

THEOREM 6.D.5: If f is a symplectic map of R^{2n} with $f(0) = 0$, and $Df(0) = I$ (I is the identity matrix), then there is a generating function $S(q, P)$ such that f is defined implicitly by the formula

$$f(q, p) = (Q, P); \quad p = \frac{\partial S(q, P)}{\partial q}, \quad Q = \frac{-\partial S(q, P)}{\partial P}$$

Proof: Let (q, p, Q, P) denote the coordinates of a point in R^{4n} and let $f(q, p) = (f_1(q, p), f_2(q, p))$. The graph K of f is a Lagrangian submanifold of R^{4n} relative to the symplectic structure $\Omega^- = dQ \wedge dP - dq \wedge dp$. The linear map $\sigma: R^{4n} \to R^{4n}; (q, p, Q, P) \to (q, P, p, -Q)$ has the property that $\sigma^* \Omega^+ = \Omega^-$, where $\Omega^+ = dQ \wedge dP + dq \wedge dp$. Therefore, $\sigma(K)$ is a Lagrangian submanifold of R^{4n} relative to the symplectic structure Ω^+. Define $\Gamma: R^{2n} \to R^{4n}; \Gamma(q, p) = (q, p, f_1(q, p), f_2(q, p))$. The range of Γ is the graph of f. Let π denote the projection of R^{4n} onto R^{2n} defined by $\pi(q, p, Q, P) = (q, p)$. Since $D\Gamma\sigma\pi(0, 0) = I$, the composite map $\Gamma\sigma\pi$ is a diffeomorphism on a neighborhood of $(0, 0)$. Therefore, the set $\sigma(K)$ can be expressed as the graph of the function $\pi\Gamma(\Gamma\sigma\pi)^{-1}$. (This function is just the inverse of the projection π restricted to $\sigma(K)$). By Theorem 6.D.1, there is a generating function S such that $\sigma(K) = \{(q, P, S_q(q, P), S_P(q, P))\}$. From the definition of σ, it follows that $K = \{(q, S_q(q, P), -S_P(q, P), P)\}$. ∎

Poincaré used a special 1-form to study the fixed points of symplectic maps. Define the 1-form β on $R^n \times R^n \times R^n \times R^n$,

$$\beta = (P - p) \, dQ - (Q - q) \, dp$$

By calculation we have $d\beta = dP \wedge dQ - dp \wedge dq$. Suppose that S is a generating function which depends only on p and Q. Then $\beta = dS$ is equivalent to the equations

$$P - p = D_Q S(p, Q) \qquad Q - q = D_p S(p, Q)$$

Alternately, the equation $\beta = dS$ determines a subset

$$F = \{(q, p, Q, P) = (Q + D_p S(p, Q), p, Q, p + D_Q S(p, Q))\}$$

We view F as the graph of a symplectic map f of $R^n \times R^n$.

PROPOSITION 6.D.6: f has a fixed point (Q, p) if and only if $dS(p, Q) = 0$.

Proof: This is obvious from the description of the set F. ∎

E. Symplectic Integrators

Symplectic integrators are symplectic maps which approximate the time s map associated with a Hamiltonian flow. The construction and testing of symplectic integrators is currently an active research area. The reader should consult Ge (1991) and Yoshida (1993) for more in-depth information on this topic.

The one-step Euler method for solving the Hamiltonian system of ordinary differential equations

$$\dot{q} = H_p(q, p) \qquad \dot{p} = -H_q(q, p) \tag{1}$$

is to fix a time step s and set

$$Q = q + sH_p(q, p) \qquad P = p - sH_q(q, p) \tag{2}$$

If $(q(t), p(t))$ is the true solution of the Hamiltonian system satisfying the initial conditions $q(0) = q, p(0) = q$, then (Q, P) approximates $q(s), p(s))$.

By Liouville's theorem, the time s map of the flow generated by the Hamiltonian system of differential equations is a symplectic map. However, map (2) above is not a symplectic map. To produce a symplectic map which is close to the Euler approximation, define a generating function $S = Pq + sH(P, q)$, and set $p = S_q$, and $Q = S_P$. Thus

$$p = P + sH_q(P, q) \qquad \text{and} \qquad Q = q + sH_P(P, q)$$

Solving, for P, we have

$$P = p - sH_q(P, q) \qquad \text{and} \qquad Q = q + sH_P(P, q)$$

This is an implicitly defined symplectic map which approximates the time s map of the flow of equations (1).

If the Hamiltonian function splits into the sum of kinetic plus potential energy and thus has the form $H = T(p) + U(q)$, then the symplectic map above is explicit.

$$P = p - sU_q(q) \qquad \text{and} \qquad Q = q + sT_P(P)$$

Exercise: Use this method to construct a symplectic integrator for the equations of a pendulum.

A new approach (Yoshida, 1993) to numerically solving systems of differential equations involves the use of Lie algebras and the Campbell–Baker–Hausdorff formula. Briefly, if X and Y are vector fields, and the flows associated with them are symbolized with the notation $\exp(tX)$, and $\exp(tY)$ respectively, then one can show that, for proper choices on constants a_i and b_i,

$$\exp(t(X+Y)) = \prod_{i=1}^{n} \exp(a_i tX) \exp(b_i tY) + O(t^{n+1})$$

This approach works well for a Hamiltonian system with Hamiltonian function or the form $H(q,p) = T(p) + U(q)$. For $n = 2$, the choice $a_i = b_i = \frac{1}{2}$ gives a second-order method. Let X and Y be the vector fields associated with the Hamiltonian functions $T(p)$ and $U(q)$, respectively. Then $X + Y$ is the vector field associated with H. The exponential formula

$$\exp(t(X+Y)) = \exp(0.5tX) \exp(tY) \exp(0.5tX) + O(t^3)$$

leads to the symplectic integration method called the *leapfrog method*:

$$q^* = q + 0.5t \operatorname{grad} T(p) \qquad P = p - t \operatorname{grad} U(q^*) \qquad Q = q + 0.5t \operatorname{grad} T(P)$$

Remark: There is also an approach to numerical integration using generating functions (Ge, 1991). One finds a power series expansion of the generating function for the time s map of the flow, and then truncates this function at some order. The symplectic map determined by the truncated generating function approximates the flow.

F. Separatrix Movement

In studying the dynamics of a parametrized family of maps, one often starts at a parameter value for which some of the dynamics is understood, and then tries to give an analysis of the dynamics for nearby parameter valves. An example is the study of periodically forced oscillations. Any one-degree-of-freedom Hamiltonian sytem of differential equations has simple dynamics. The solutions are contained in the constant level curves of the Hamiltonian. To be explicit, consider the Hamiltonian

$$A(x, y) + \varepsilon B(x, y, t, \varepsilon)$$

where B is periodic of period 1 in t. The variables x and y are real, and ε is a parameter. Associate a one-parameter family of maps $F(x, y; \varepsilon)$ of the plane by defining

$$F(x_0, y_0; \varepsilon) = (x(1), y(1))$$

where $(x(t), y(t), t)$ is the solution to the initial value problem

$$\dot{x} = \frac{\partial}{\partial y}\{A(x,y) + \varepsilon B(x,y,t,\varepsilon)\}, \qquad \dot{y} = -\frac{\partial}{\partial x}\{A(x,y) + (x,y,t,\varepsilon), \dot{t} = 1$$

$$x(0) = x_0, \qquad y(0) = y_0, \qquad t(0) = 0$$

Assume that the functions A and B are smooth, so that solutions exist, and are unique. Then the family of maps preserves area. The map F_0 has the function A as an integral, and therefore orbits of this map are contained in the constant level curves of A. Assume that there is one level curve of A that has the form of a fish (Figure 6.1).

We assume that the point p is a hyperbolic fixed point of the map F_0, and the loop through p consists of homoclinic points. The question is what happens to this structure when the map is perturbed. Typically, the loop coils up into a homoclinic tangle of stable and unstable manifolds lacing through a sequence of transverse homoclinic points. A method due to Poincaré, Arnold, and Melnikov, allows one to compute the velocity of movement of stable and unstable manifolds as the parameter varies, and consequently to establish the existence of transverse homoclinic points.

Choose a vector L which is transverse to the unstable manifold of p at the point q. The hyperbolic fixed point p persists for some open interval of values of the parameter ε. Let $p(\varepsilon)$ be the continued hyperbolic point. The unstable manifold of $p(\varepsilon)$ is close to the unstable manifold of p, and therefore intersects the line spanned by L transversely at a point $q(\varepsilon) = s(\varepsilon)L$ which is close to q as pictured in Figure 6.2. The *displacement vector* at q is defined to be the vector

$$s'(0)L$$

Notice that the displacement vector describes the rate of change with respect to the unstable manifold at the point q. By varying the point q, and the vector L, the velocity of movement can be calculated all along the unstable manifold. A standard choice of $L = L(q)$ is to choose it as a unit vector perpendicular to the unstable manifold at q. Similarly, a displacement vector is defined for the movement of the stable manifold of p, and is denoted $V^s(q, L)$. A family $q^*(\varepsilon)$ of transverse homoclinc points arises at a point q^* on the homoclinic loop where the difference of the displacement vector fields has a nondegenerate zero.

The displacement vectors can be calculated as follows: Choose a unit vector v_0 tangent to the unstable manifold W_0^u at a point q_0. Define

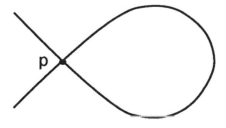

Figure 6.1. Level curve in the form of a fish.

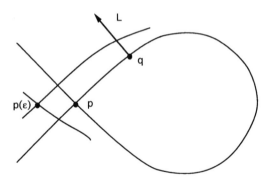

Figure 6.2. Movement of stable and unstable manifolds. The stable manifold of $p(\varepsilon)$ intersects the segment L in a point. The velocity of this intersection point when $\varepsilon = 0$ is the displacement vector field at the point q.

$q_k = F_0^{-k}(q_0)$ and define $v_k = DF_0^{-k}(q_0)(v_0)$. Also define $q_0(\varepsilon)$ to be the "first" intersection of W_ε^u with the line spanned by L. (W_ε^u denotes the unstable manifold of $p(\varepsilon)$ with respect to the map F_ε.)

By the stable manifold theorem, W_ε^u depends smoothly on ε and hence the intersection point $q_0(\varepsilon)$ depends smoothly of ε. Define $q_k(\varepsilon) = F_\varepsilon^{-k}(q_0(\varepsilon))$. Define $\Delta(\varepsilon) = \Omega(q_0(\varepsilon) - q_0, v_0)$.

THEOREM 6.F.1: Assume that the family of maps has the form $F_\varepsilon(z) = F_0(z) + \varepsilon F_1(z, \varepsilon) + \varepsilon^2 F_2(z, \varepsilon)$, where the functions F_0, F_1, F_2 are smooth. If differentiation with respect to ε is denoted by a prime, and we define $w_k = q_k'(0)$, then

$$\Omega(w_0, v_0) = \sum_{j=1}^{\infty} \Omega(F_1(q_j, 0), v_{j-1})$$

Proof: From the definitions, $q_0(\varepsilon) = F_0(q_1(\varepsilon)) + \varepsilon F_1(q_1(\varepsilon), \varepsilon) + \varepsilon^2 F_2(\varepsilon), \varepsilon)$. Thus $w_0 = DF_0(q_1)w_1 + F_1(q_1, 0)$. Since $v_0 = DF_0(q_1)v_1$, it follows that

$$\Omega(w_0, v_0) = \Omega(DF_0(q_1)w_1, DF_0(q_1)v_1) + \Omega(F_1(q_1, 0), v_0)$$
$$= \Omega(w_1, v_1) + \Omega(F_1(q_1, 0), v_0)$$

However, $\Omega(w_1, v_1)$ can be replaced by a similar expression. Continuing, for n steps we have

$$\Omega(w_0, v_0) = \Omega(w_{n+1}, v_{n+1}) + \sum_{j=1}^{n} \Omega(F_1(q_j, 0), v_{j-1})$$

The vectors w_n are bounded as $n \to -\infty$ because the unstable manifold depends smoothly on ε. The norms of the vectors $|v_n| \to 0$ as $n \to -\infty$ because v_n is tangent to the unstable manifold W_0^u and DF_0^{-1} compresses such vetors. Therefore,

$$\Omega(w_0, v_0) = \sum_{j=1}^{\infty} \Omega(F_1(q_j, 0), v_{j-1})$$

■

G. Normal Forms

Suppose that F is a real analytic symplectic map of $(R^{2n}, dq \wedge dp)$ such that $F(0) = 0$ and $M = DF(0)$. We expand F in a Taylor series about the origin. Thus

$$F(z) = Mz + F_2(z) + F_3(z) + \ldots$$

where $F_d(z)$ denotes terms of degree d.

The goal is to make a symplectic change of coordinates such that the resulting map F^* can be factored into the product of n two-dimensional maps plus a small perturbation.

Definition: Assume that a symplectic matrix M has distinct eigenvalues $\lambda_1, \ldots, \lambda_n, \lambda_1^{-1}, \ldots, \lambda_n^{-1}$. Let the symbol α denote a vector of nonnegative integers $\alpha = (\alpha_1, \ldots, \alpha_n)$, and define $|\alpha| = \alpha_1 + \cdots + \alpha_n$. Define $\lambda^\alpha = (\lambda_1^{\alpha_1} \cdots \lambda_n^{\alpha_n})$. The matrix M satisfies *nonresonance conditions* up to order r provided $(\lambda^\alpha - 1) \neq 0$ whenever $|\alpha| \leq r$.

The following result is called the Birkhoff normal form for a symplectic map. Our presentation of this result is adapted from a paper by Moser (1977).

THEOREM 6.G.1: Let F be an analytic symplectic map of R^{2n} which has the origin as a fixed point. Suppose that the linear part M of F has distinct eigenvalues which satisfy nonresonance conditions up to order $d + 1$. Then there exists a symplectic change of variables h such that the map $F^* = h^{-1}Fh$ has the symmetry $F^*M = MF^*$ (up to terms of order d). If $|\mu| = 1$ for each eigenvalue of M, then F^* has the form $F^*(q, p) = (Q, P)$ with

$$Q_k = q_k \cos(\theta_k) - p_k \sin(\theta_k) + O(d+1)$$
$$P_k = q_k \sin(\theta_k) + p_k \cos(\theta_k) + O(d+1)$$
$$\theta_k = \alpha_k + \beta_k(r_1, \ldots, r_n) \quad r_k^2 = (q_k^2 + p_k^2) \quad \lambda_k = e^{i\alpha_k}$$

The function B_k is a power series with constant term equal to zero.

The proof of this theorem will occupy the rest of this section.

Let $h(z)$ be a symplectic change of coordinates which is close to the identity map. Thus $h(z = z + h_d(z) + O(d + 1)$, where $h_d(z)$ denotes terms of degree d in the Taylor expansion of h and $O(d + 1)$ denotes the higher order terms. The new map F^* conjugate to F is $F^* = h^{-1}Fh$. Alternately,

$$Fh = hF^* \quad (1)$$

An important point to notice is that the map F can be written $F = Mf$ with $f = M^{-1}F$. Thus, f is a symplectic map with $Df(0)$ equal to the identity matrix I. Similarly, one can write $F^* = Mf^*$ with $Df^*(0) = I$.

We set $z = (q, p)$, and let (q, p, Q, P) be coordinates on R^{4n}. First we will make a general change coordinates h, and see what happens. Then we will determine the "right" change of coordinates to put F^* in the form specified by the theorem.

LEMMA 6.G.2: Let g be a real analytic symplectic map of R^{2n} of the form $g(z) = z + g_d(z) + O(d + 1)$. Then there is a real analytic function G such that $g_d(z) = J \operatorname{grad} G(z)$.

Proof: Let $z = (q, p)$, and let $g(z) = (g_1(z), g_2(z))$. The 1-form $\beta = (P - p)\,dQ - (Q - q)\,dp$ is exact on the graph of g because g is symplectic. Therefore, there is a real analytic function $W(q, p, Q, P)$ defined on a neighborhood of the origin in R^{4n} such that $\beta - dW = 0$ on the graph of g. It follows that the function W depends only on the variables Q and p. Further, it follows that

$$g(q, p) = (q, p) + (W_p(g_1(q, p), p), -W_Q(g_1(q, p))p)$$

The function $G(q, p) = W_d(q, p)$ is the required function. ∎

We assume by induction that coordinate changes have been applied so that $f(z) = z + f_d(z) + O(d + 1)$. Apply Lemma 6.G.2 to write

$$f(z) = z + JS_z(z) + O(d + 1)$$
$$h(z) = z + JV_z(z) + O(d + 1)$$
$$f^*(z) = z + JS^*_z(z) + O(d + 1)$$

where d is greater than or equal to 3, and the functions S, V, and S^* are homogeneous of degree $d + 1$.

LEMMA 6.G.3: If $G(z)$ is a homogeneous polynomial of degree d and $DG(z) = 0$ for all z, then $G = 0$.

Proof: This is an exercise.

LEMMA 6.G.4: The generating functions V, S, and S^* satisfy the equation

$$S^*(z) = S(z) + V(z) - V(Mz)$$

Proof:

$$\begin{aligned}
fh(z) &= f(z + JV_z(z) + O(d+1)) \\
&= z + JV_z(z) + JS_z(z + JV_z(z) + O(d+1))) + O(d+1)) \\
&= z + J[V_z(z) + S_z(z)] + O(d+1) \\
Fh(z) &= Mz + MJ[V_z(z) + S_z(z)] + O(d+1) \\
F^*(z) &= Mf^*(z) = Mz + MJS_z^*(z) + O(d+1) \\
hF^*(z) &= F^*(z) + JV_z(F^*(z)) + O(d+1) \\
&= Mz + MJS_z^*(z) + JV_z^*(Mz) + O(d+1)
\end{aligned}$$

Since the matirx M is symplectic, we have $M^t JM = J$. Note that $JM^t = -JM^t JJ = M^{-1}J$, and from this we derive the identity $MJM^t = J$. Therefore,

$$hF^*(z) = Mz + MJ[S_z^*(z) + M^t V_z(Mz)] + O(d+1)$$

In order for the equation $Fh = hF^*$ to be satisfied, the homogeneous parts of degree d of both sides must be equal. Hence we have

$$\frac{\partial}{\partial z}[V(z) + S(z)] = \frac{\partial}{\partial z} S^*(z) + M^t \frac{\partial}{\partial z} V(Mz) = \frac{\partial}{\partial z}[S^*(z) + V(Mz)]$$

Because V, S and S^* are homogeneous, Lemma 6.G.3 implies that

$$S^*(z) = S(z) + V(z) - V(Mz) \tag{2}$$

Our strategy is to choose the function V so that the function S^* generates simple dynamics. To use equation (2) it is necessary to compute $V(Mz)$. This computation proceeds according to the following steps.

Step 1. Choose a complex symplectic matrix A so that $A^{-1}MA = \Lambda$, with

$$\Lambda = \text{diag}(\lambda_1, \ldots, \lambda_n, \lambda_1^{-1}, \ldots, \lambda_n^{-1})$$

Step 2. Let $z = A\zeta$, and define $W(\zeta) = V(z)$. Then

$$W(\Lambda\zeta) = W(A^{-1}MA\zeta) = W(A^{-1}Mz) = V(Mz)$$

Step 3. W is homogeneous of degree d since V is homogeneous of degree d.

$$W(\zeta) = \sum W_\alpha \zeta^\alpha \quad \text{where} \quad \alpha = (\alpha_1, \ldots, \alpha_{2n}), \zeta^\alpha = \prod_{k=1}^{2n}(\zeta_k)^{\alpha_k}$$

Step 4. Compute $V(Mz)$:

$$V(Mz) = W(\Lambda \zeta) = \sum W_\alpha \lambda^\alpha \zeta^\alpha \quad \text{where} \quad \lambda^\alpha = \lambda_1^{\alpha_1 - \alpha_{1+n}} \cdots \lambda_n^{\alpha_n - \alpha_{2n}}$$

Therefore,

$$V(z) - V(Mz) = \sum (1 - \lambda^\alpha) W_\alpha \zeta^\alpha$$

Now we impose a symmetry condition that makes S^* generate simple dynamics. We require that $S^*(Mz) = S^*(z)$. The symplectic map generated by S^* will then have the symmetry $F^* M = MF^*$. This leaves the important set of equations to be solved:

$$S^*(Mz) = S^*(z) \quad V(Mz) - V(z) = S^*(z) - S(z) \tag{3}$$

As with the function V, we (uniquely) determine functions T and T^* on C^{2n} so that $TA = S$ and $T^*A = S^*$. These functions are expressed as power series in the same manner as W.

The equations (3) can now be written in terms of the coefficients of the various series as

$$(\lambda^\alpha - 1)T_\alpha^* = 0 \quad (\lambda^\alpha - 1)W_\alpha = T_\alpha^* - T_\alpha \tag{4}$$

If $(\lambda^\alpha - 1) = 0$, then a solution is $T_\alpha^* = T_\alpha$; $W_{\alpha\alpha} = 0$. If $(\lambda^\alpha - 1) \neq 0$, then the solution is $T_\alpha^* = 0$, $W_\alpha = (1 - \lambda^\alpha)^{-1} T_\alpha$.

Proof of Theorem 6.G.1: The argument above shows that equations (3) can be solved when there are no resonances. Now suppose that the matrix M has eigenvalues on the unit circle. Then M can be put in block diagonal form with the diagonal blocks being 2×2 rotation matrices. Equivalently, the eigenspace corresponding to the pair of eigenvalues λ_j, and λ_j^{-1} is a two-dimensional symplectic subspace of R^{2n}. The action of M is to rotate this plane through an irrational multiple of 2π. Therefore, the symmetry $S^*(Mz) = S^*(z)$ now implies that the function S^* must depend only on the radii and not the angles in these rotation planes. Therefore, the transformation generated has the form asserted in the theorem. ■

The normal form process produces a coordinate system in which low-order terms of the map are simple. However, the size of the region where higher order terms do not contribute to the dynamics may be exceedingly small. The structure of orbits near an elliptic fixed point in four or more dimensions is not understood. There is no proof that such a fixed point is stable except in the case where the matrix M is positive or negative definite.

H. Problems

1. Let L be a Lagrangian subspace of V^{2n}. Show that there is a symplectic map which transforms L to the n-plane of vectors whose last n entries are zero.
2. Show that the graph of a linear symplectic transformations of V^{2n} is a Lagrangian plane in V^{4n}.

I. Further Reading

The literature on symplectic dynamics and symplectic geometry is growing rapidly. Classic references are Siegel and Moser (1971), Arnol'd (1978), Abraham and Marsden (1978). The book of Meyer and Hall (1991) is an excellent reference on Hamiltonian dynamics. A reprint collection of some important research papers is found in MacKay and Meiss (1987). For recent advances in symplectic geometry, see Hofer and Zehnder (1994). There is some interesting material on normal forms for maps in Arrowsmith and Place (1990). I recommend the papers on symplectic integrators by Ge (1991) and Yoshida (1993). The references in each of these books and articles may be consulted to enlarge this very brief guide to the literature.

7

Invariant Measures

A. Measure Spaces

A measure is a generalization of what we mean by area in the plane and volume in three dimensions. A measure assigns a "size" to a set. Thus, for example, the measure of disjoint sets should be the sum of the measures of each. It turns out that there are some sets so "bad" that they cannot be assigned a measure in a natural way. The collection of "good" sets forms what is called a sigma algebra and a measure assigns a nonnegative number to each set in the sigma algebra. Measure and integration theory are standard topics taught in first-year graduate mathematics courses in analysis. Some definitions that are needed in the next section are given below for reference.

The reader who is not familiar with measure theory may just skip this section and keep in mind that good sets in a Euclidean space are approximated by disjoint unions of generalized boxes, and that the volume of a box is the product of the lengths of its edges.

Definitions: A *sigma algebra* of subsets of a set X is a family $\mathcal{A}$ of subsets of X such that

> X belongs to $\mathcal{A}$.
> The complements of sets in $\mathcal{A}$ belong to $\mathcal{A}$.
> The countable union of sets in $\mathcal{A}$ belongs to $\mathcal{A}$.

The set of all subsets of X obviously forms a sigma algebra. The intersection of a collection of sigma algebras is a sigma algebra (exercise).

For a metric space (X, d), the *Borel sigma algebra* is the intersection of all sigma algebras which contain all open subsets of X. Sets in the Borel sigma algebra are called *Borel sets*.

A *measure space* $(X, \mathcal{A}, \mu)$ consists of a set X, a sigma algebra $\mathcal{A}$ of subsets of X, and a function $\mu: \mathcal{A} \to [0, \infty]$ such that for any sequence $\{A_j\}$ of disjoint sets

$$\mu(\cup \{A_j : 1 \leq j < \infty\}) = \sum_{j=1}^{\infty} \mu(A_j)$$

The function μ is called a *measure* on X. The function μ is called a *Borel measure* on X if the sigma algebra is the Borel sigma algebra.

The set of all Borel measures on X is denoted by $\mathcal{M}(X)$. This set is convex since for $0 \leq t \leq 1$, the combination of measures $t\mu + (1-t)\nu$ is a measure.

B. Invariant Measures

Suppose that f is a map of X. One can show that if E is a Borel set then, because f is continuous, the set, $f^{-1}(E)$ is also a Borel set. The function f induces a transformation f^* of $\mathcal{M}(X)$. If μ is a measure then a new measure $f^*\mu$ is defined by $f^*\mu(E) = \mu(f^{-1}(E))$.

An *f-invariant measure* is a measure μ such that $f^*\mu = \mu$. The invariant measures are the fixed points of the induced dynamics on the space of measures. It is easy to check that the set of f-invariant Borel measures is convex.

For a symplectic map, there is an invariant measure defined by integrating the symplectic volume form. If an invariant measure exists for a general map, then this has the following consequences for the dynamics.

THEOREM 7.B.1: Suppose that μ is an invariant Borel measure for a map f. Suppose that the map f has an inverse. Then given a Borel set E such that $\mu(E)$ is finite, the set of orbits which enter E and do not exit is a Borel set and has measure zero.

Proof: Since the function f has an inverse, the backward exit time from E is well defined. Let e_j denote the set of points of E with backward exit time j. Then these sets are disjoint Borel sets and, further, $f(e_j) = e_{j+1}$. The measures of all sets e_j are equal because f preserves the measure μ. Since the measure of E is finite, the measure of each set e_j must be zero. ∎

Remark: This theorem can be used to show that according to the Newtonian model of celestial mechanics the probability of a new planet coming from outer space and being captured by our solar system is zero.

COROLLARY 7.B.2 (The Poincaré Recurrence Theorem): Suppose that μ is an invariant Borel measure for an invertible map f and suppose that the measure of X is finite. If U is a Borel set of positive measure, then almost every point in U returns to U infinitely often under iteration by f.

Proof: Let $E = X - U$ and let U_1 denote the set of points in U whose iterates never return to U. By Theorem 7.B.1, this set has measure zero. Now consider the map $g = f^k$. The function g also preserves the measure. Let U_k

denote the set of points in U whose g-iterates never return to U. Again, the set U_k has measure zero. Thus the union of the sets U_k for all k has measure zero. Each point not in this union has an iterate which for some multiple of k returns to U. Since k is arbitrary, each such point has infinitely many iterates which return to U. ∎

COROLLARY 7.B.3: If the measure of X is finite, and if each open set has positive measure, and if X is path connected, then the chain recurrent set of f is all of X, and one can chain between any pair of points.

Proof: Let $\varepsilon > 0$ be given. Choose an ε-neighborhood U of a point x. Choose a point y which returns to U at the kth iterate, and which is so close to x that $f(x)$ and $f^k(y)$ are distance less than ε apart. Then the chain consisting of $x, f(y), \ldots, f^{k-1}(y), x$ is an ε-chain from x to x. Further, with a 2ε-chain one can chain from x to any point in an ε-neighborhood of the point x. To chain from x to another point z, since X is path connected, one can choose a sequence of points $\{z_j\}$ starting at x, and ending at z with the distance between each successive pair less than ε. Fill in the "gap" from z_j to z_{j+1} with a 2ε-chain. Thus one can 2ε-chain from x to z. Since ε is arbitrary, one can ε-chain between any pair of points. ∎

In the case where the space X is compact, and has an f-invariant probability measure μ, a fundamental theorem due to Birkhoff gives important statistical information about where orbits spend their time.

For a set E let χ_E denote the *characteristic function* of the set E which takes the value 1 on E and the value 0 on the complement of E. Define the function

$$\text{Ave}(x, E, n) = \frac{1}{n} \sum_{j=0}^{n-1} \chi_E(f^j(x))$$

The average time that the orbit of an initial point x spends in E (if it exists) is

$$\text{Ave}(x, E) = \lim_{n \to \infty} \text{Ave}(x, E, n)$$

It is common practice to use the term *almost everywhere* to mean everywhere except for a set of measure zero.

THEOREM 7.B.4: For a Borel set E, the limit $\text{Ave}(x, E)$ exists for almost every x. Furthermore, $\text{Ave}(x, E)$ is an integrable function of x.

Proof: Define $\overline{\text{Ave}}(x, E) = \limsup A(x, E, n)$ and $\widetilde{\text{Ave}}(x, E) = \liminf A(x, E, n)$. A basic result from the theory of measure and integration states that if the integral of a nonnegative function is zero, then the function must be zero almost everywhere. Thus it is sufficient to show that $\int_x \overline{\text{Ave}}(x, E) - \widetilde{\text{Ave}}(x, E) = 0$.

Define $T(x, E, \varepsilon)$ to be the least integer n such that $\text{Ave}(x, E, n) \geq \overline{\text{Ave}}(x, E) - \varepsilon$. Further, define $B(x, E, k) = \{x : T(x, E, \varepsilon) > k\}$. These sets are measurable. Further, the measures of the sets $B(x, E, k)$ must converge to zero as k approaches infinity: If not, the intersection of these sets has positive measure. But this is impossible, because for a point y in this intersection, one would have $\text{Ave}(y, E, n) < \overline{\text{Ave}}(y, E) - \varepsilon$ for all integers n, thus contradicting the definition of $\overline{\text{Ave}}(y, E)$.

LEMMA 7.B.5: $\int_x \overline{\text{Ave}}(x, E) \leq \mu(E)$

Proof: From the definition, and the fact that μ is an invariant measure, it follows that $\int_x \text{Ave}(X, E, n) = \mu(E)$. (To see this start with $n = 1$ and $n = 2$.)

Choose k sufficiently large so that $\mu(B(x, E, k)) < \varepsilon$. Define $E' = \{T(x, E, \varepsilon) > k\}$ and define $E'' = E \cup E'$.

We will show that $\text{Ave}(x, E'', n) \geq (1 - k/n)[\overline{\text{Ave}}(x, E'') - \varepsilon]$. To see this, break this orbit of x, into segments of as follows: If x belongs to E', form the segment which includes all successive points which remain in E'. If x does not belong to E', choose the segment to be of length k. Continue this process, with a final segment of length less than or equal to k. Average the number of visits to the set E'' over each of these segments. The first type segment has average 1; the second type has average greater than or equal to $\mu(E) - \varepsilon$; and the last segment has average zero at worst. Now do the arithmetic, and this proves the inequality.

Next integrate this inequality to get $\mu(E'') \geq (1 - k/n)[\int_x \overline{\text{Ave}}(x, E) - \varepsilon]$. Let n go to infinity to get the inequality $\mu(E'') \geq \int_x \overline{\text{Ave}}(x, E) - \varepsilon$. Since $\mu(E') \leq \varepsilon$, it follows that $\mu(E) \geq \int_x \overline{\text{Ave}}(x, E) - 2\varepsilon$. Now let ε go to zero to finish the proof. ∎

LEMMA 7.B.6: $\int_x \widetilde{\text{Ave}}(x, E) \geq \mu(E)$.

The proof of this lemma is analogous to the preceding proof. Putting the two lemmas together, we have shown that $\int_x \overline{\text{Ave}}(x, E) - \widetilde{\text{Ave}}(x, E) = 0$. ∎

Definition: A map f is *ergodic* with respect to an invariant measure μ if, given a Borel set E, then $\text{Ave}(x, E) = \mu(E)$ for almost every x.

It is a difficult problem to determine whether a pair is ergodic. It is even difficult sometimes to show that invariant measures exists for a given map. Sometimes f-invariant measures can be produced by averaging the translates of a given measure. If f is a map of a metric space X and μ is a Borel measure on X, then sometimes there exists an invariant Borel measure ν on X such that for each Borel set E,

$$v(E) = \lim_{n\to\infty} \frac{1}{n} \sum_{j=0}^{n-1} (f^j)^* \mu(E)$$

However, it is not clear that this limit exists. Note that the averaged measure may not necessarily be positive on open sets.

Open Research Question: For the standard map family, are there initial conditions in the numerically determined ergodic zones whose omega limit sets have positive area? The computer picture in Figure 1.7 seems to suggest that this is the case.

C. Further Reading

The literature on ergodic theory is extensive. A place to start to learn more about this field is with the books by Billingsley (1965), Walters (1981), and Mane (1987), and the references cited by their authors.

Appendix A: Metric Spaces

Metric spaces will be used to represent the sets of states of physical systems. A metric measures the "distance" between two states. In Euclidian space R^n the distance between points x and y is defined by the formula

$$d(x, y) = \left(\sum_{j=1}^{n} (x_j - y_j)^2 \right)^{1/2}$$

The definition is motivated by the Pythagorean theorem, which states that the square of the hypotenuse of a right triangle is the sum of the squares of its sides. Metrics generalize the key properties of the Euclidian distance function including the triangle inequality, which states that the distance from x to z is less that or equal to the sum of the distances from x to y and from y to z. The reader is encouraged to prove the propositions stated below or to consult topology textbooks as needed.

A. Definitions

A *metric space* is a pair (X, d) consisting of a set X and a function $d: X \times X \to [0, \infty)$ called a *metric* such that

(1) $d(x, y) = 0$ if and only if $x = y$.
(2) $d(x, y) = d(y, x)$.
(3) $d(x, y) + d(y, z) \leq d(x, z)$.

A *sequence* s is a function from the set of nonnegative integers Z^+ to X. The nth point of the sequence is the point $s(n)$. Sometimes, when the sequence is understood, we set $x_n = s(n)$.

A *Cauchy sequence* is a sequence $\{x_n\}$ such that given $\varepsilon > 0$ there exists a positive integer k such that $d(x_n, x_m) < \varepsilon$ whenever n and m are greater than k.

A *subsequence* of a sequence s is any sequence σ of the form $\sigma(j) = \sigma(\alpha(j))$, where α is a strictly increasing function from Z^+ to Z^+.

A sequence of points $\{x_n\}$ in X *converges* to a point y in X if given $\varepsilon > 0$ there exists an integer m such that $d(x_n, y) < \varepsilon$ for all $n \geq m$.

A metric space X is *complete* if every Cauchy sequence in X converges to a point in X.

A point p is a *limit point* of a subset Y of X provided that there is a sequence of points in Y which converges to p.

A subset Y of X is *closed* if it contains all its limit points. The *closure* of Y denoted by cl(Y) is the set of all limit points of Y.

The *ball* of radius r centered at x is the set $B(x, r) = \{y, \in X : d(x, y) < r\}$.

A subset U of X is *open* if for each point x belonging to U there exists $r > 0$ such that the ball $B(x, r)$ is contained in U. The radius r may depend on x.

PROPOSITION A.1

(1) If Y is a closed set then $X - Y$ is an open set.
(2) If U is an open set then $X - U$ is a closed set.

Suppose that X and Y are metric spaces. A function f from X to Y is *continuous* if for every open subset V of Y the set $f^{-1}(V)$ is an open subset of X.

f is a *homeomorphism* if f is continuous and invertible, and the inverse of f is continuous.

If there is a homeomorphism between two spaces, then the spaces are the same from a topologist's point of view. A central problem in topology is to decide when two spaces are homeomorphic.

PROPOSITION A.2: A function f from X to Y is continuous if and only if given any point x in X and any sequence $\{x_n\}$ converging to x then the sequence $\{f(x_n)\}$ converges to $f(x)$.

PROPOSITION A.3: If f is continuous and A is a subset of X, then

$$f(\text{cl}(A)) \subset \text{cl}(f(A))$$

Definitions: The *interior* of a subset S of a metric space X is the set defined by

$$\text{int}(S) = X - \text{cl}(X - S)$$

A *neighborhood* of a point x in a metric space X is a subset N such that x belongs to the interior of N.

The *exterior* of a subset S of a metric space X is the set

$$\text{ext}(S) = X - \text{cl}(S)$$

APPENDIX A 133

The *boundary* of a subset S of a metric space X is the set

$$\partial S = \text{cl}(S) \cap \text{cl}(X - S)$$

A subset S of a metric space X is *disconnected* if there exist disjoint open sets U and V each having nonempty intersection with S such that S is contained in the union of U and V.

A subset S of a metric space X is *connected* if it is not disconnected.

PROPOSITION A.4: If N is a subset of a metric space X and S is a connected subset of X whose intersection with both the interior and the exterior of N is nonempty, then the intersection of S with the boundary of N is nonempty.

Definitions: A subset K of a metric space X is *compact* if every sequence of points in K has a convergent subsequence.

A metric space X is *locally compact* if each point x in X has a neighborhood which is compact.

A subset K of a metric space X is *bicompact* if given any collection of open sets whose union contains K there exists a finite number of these sets whose union also contains K.

PROPOSITION A.5: A subset K of a metric space X is compact if and only if it is bicompact.

PROPOSITION A.6: Suppose that f is a continuous function from a metric space X to a metric space Y. Suppose that C is a connected subset of X and K is a compact subset of X. Then

(1) $f(C)$ is connected.
(2) $f(K)$ is compact.

PROPOSITION A.7: Suppose that g is a continuous real-valued function defined on a compact subset K of a metric space. Then there exist points x and y in K such that $g(K)$ is contained in the interval $[g(x), g(y)]$. Thus g achieves its maximum and its minimum on K.

A function f from a metric space (X, d) to a metric space (Y, D) is *uniformly continuous* if given $\varepsilon > 0$ there exists $\delta > 0$ such that $d(x_1, x_2) < \delta$ implies that $D(f(x_1), f(x_2)) < \varepsilon$.

PROPOSITION A.8: Suppose that f is a continuous function from a compact metric space X to a metric space Y. Then f is uniformly continuous.

PROPOSITION A.9. If $\{K_j\}$ is a nested sequence of compact sets and K denotes the intersection of these sets, then K is nonempty. Furthermore, if U

is a neighborhood of K, then there exists a positive integer m such that K_j is contained in U for all $j > m$.

PROPOSITION A.10: Suppose that X and Y are compact metric spaces. Let $f: X \to Y$ be continuous, one-to-one, and onto. Then the inverse of f is continuous and hence f is a homeomorphism.

There is a particularly interesting metric space called the Cantor set which often occurs in the study of dynamics.

Example: The *standard Cantor set* is a compact subset of the unit interval [0, 1] which is constructed in a sequence of stages. At the first stage, the middle third (1/3, 2/3) of the unit interval is removed. At the second stage, the middle thirds (1/9, 2/9) and (5/9, 6/9) of the remaining two intervals are removed. At the nth stage, 2^n open intervals forming the middle thirds of the closed intervals from the previous stage are removed. The standard Cantor set is the set of points which are never removed. Another way of defining this set is

$$C = \left\{ \sum_{k=1}^{\infty} a_k 3^{-k} : a_k = 0 \text{ or } a_k = 2 \right\}$$

Metric spaces which are homeomorphic to the standard Cantor set have been characterized by topological properties: They are compact; and each point in the space has other points of the space arbitrarily close to it; and the only connected subsets of the space consist of single points. Thus, in the topologists' jargon, Cantor sets are compact, perfect, totally disconnected metric spaces. It is an informative exercise to prove that the standard Cantor set has these properties. A fairly difficult exercise is to show that any two compact, perfect, totally disconnected metric spaces are homeomorphic. A classic result in point set topology is the Hahn–Mazurkiewicz theorem, which says that the standard Cantor set can be mapped continuously onto any compact metric space.

B. The Hausdorff Metric

Let S be a subset of a metric space X and let x be a point of X. Then the *distance* between x and S is the number

$$d(x, S) = \inf \{d(x, y): y \in S\}$$

A sequence of points $\{x_n\}$ in X *converges* to a subset S of X if

$$d(x_n, S) \to 0 \quad \text{as } n \to \infty$$

Let A and B be compact subsets of a metric space X. Let $A(r)$ denote the set of all points of distance less than r from A. Then the *distance* between A and B is the number

$$d(A, B) = \inf \{r \geq 0 \colon B \subset A(r) \text{ and } A \subset B(r)\}$$

Let $H(X)$ denote the set of all compact subsets of a complete metric space X. Define the Hausdorff metric $d(A, B)$ between points A and B of $H(X)$ as above.

PROPOSITION A.11: $H(X)$ together with the Huasdorff metric is a complete metric space.

A map g of a metric space X is a *contraction map* if there exists a constant K between 0 and 1 such that $d(g(x_1), g(x_2)) \leq Kd(x_1, x_2)$ for all points x_1 and x_2.

PROPOSITION A.12: If g is a contraction map on a complete metric space X, then there exists a unique "fixed point" p such that $g(p) = p$. Furthermore, for any x belonging to X, the sequence $g^n(x)$ converges to p.

Suppose g is a contraction map on X. Then there is an induced map G on $H(X)$ defined by $G(A) = \{g(x) \colon x \in A\}$.

PROPOSITION A.13: With the above conditions, G is a contraction map with contraction constant K.

THEOREM A.14: Suppose that for $1 \leq n \leq N$, g_n is a contraction map. Define $G \colon H(X) \to H(X)$ by

$$G(A) = \cup \{g_n(A) \colon 1 \leq n \leq N\}$$

Then G is a contraction map on $H(X)$.

C. Fractals

Suppose that G is a contraction map which satisfies the conditions of Theorem A.14. The unique fixed point Z of G is "self-similar" or "fractal" in the sense that it is the union of its images with respect to the N contraction maps g_n.

Example: (The Koch curve). Let E be the interval $[-1, 1]$ on the x-axis in the plane. For $1 \leq j \leq 4$ let g_j be contraction maps of the plane defined as follows: Each map is the composition of the contraction map c, with a rotation and then a translation. The map c is the linear contraction

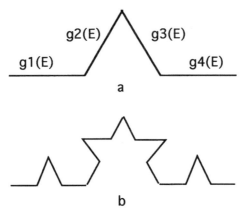

Figure A.1. Two steps in the construction of a fractal called the Koch curve.

$c: (x, y) \to \frac{1}{3}(x, y)$. The rotations and translations are chosen so that the sets $g_j(E)$ are as pictured in Figure A.1a. Thus, the associated map G on $H(R^2)$ maps E to a polygonal line with four line segments. Figure A.1b shows $G^2(E)$. The unique fixed point of G is the Koch curve.

Appendix B: Numerical Methods for Ordinary Differential Equations

Let $F: R^n \to R^n$ be a function with continuous partial derivatives up to some order consistent with differentiability hypotheses needed to make sense of the arguments to follow. We suppose that $y(t)$ solves the initial value problem $y' = F(y)$, $y(0) = y_0$, and we set

$$t_j = jh \qquad y_j = y(t_j) \qquad \text{and} \qquad y'_j = y'(t_j)$$

for each integer j. From the fundamental theroem of calculus,

$$y_{n+1} = y_{n-1} + \int_{t_{n-1}}^{t_{n+1}} y'(t)\, dt = y_{n-1} + \int_{t_{n-1}}^{t_{n+1}} F(y(t))\, dt$$

Approximating the integral in this expression by $2hF(y_n)$, one obains the *midpoint rule*:

$$u_{n+1} = u_{n-1} + 2hF(u_n)$$

An alternate derivation of the midpoint rule uses Taylor's theorem:

$$y(t+h) = y(t) + y'(t)h + y''(t)\left[\frac{h^2}{2}\right] + y'''(c)\left[\frac{h^3}{3!}\right] \qquad \text{for some } c \in [t, t+h]$$

$$y(t-h) = y(t) - y'(t)h + y''(t)\left[\frac{h^2}{2}\right] - y'''(d)\left[\frac{h^3}{3!}\right] \qquad \text{for some } d \in [t, t-h]$$

Subtract the second equation from the first and solve for $y'(t)$:

$$y'(t) = \frac{[y(t+h) - y(t-h)]}{[2h]} + [y'''(c) + y'''(d)]\left[\frac{h^3}{3!}\right]$$

Since $y'(t) = F(y(t))$, it follows that

$$y(t+h) = y(t-h) + 2hF(y(t)) - [y'''(c) + y'''(d)]\left[\frac{h^3}{3}\right]$$

Therefore, the midpoint method has local truncation error $O(h^3)$ provided that the third derivative of $y(t)$ is bounded.

Again using the fundamental theorem of calculus,

$$y_{n+1} = y_n + \int_{t_n}^{t_{n+1}} y'(t)\,dt = y_n + \int_{t_n}^{t_{n+1}} F(y(t))\,dt$$

Using the trapezoid rule to approximate the integral, one obtains the *trapezoid method*:

$$u_{n+1} = u_n + \left[\frac{h}{2}\right][F(u_n) + F(u_{n+1})]$$

Notice that this method is an implicit method because u_{n+1} appears on both sides of the equation. To compute with this method one might "predict" the vector u_{n+1} using the Euler method ($u_{n+1} = u_n + hF(u_n)$) and then "correct" using the trapezoid method to get the *Heun method*:

$$v_{n+1} = u_n + hF(u_n)$$
$$u_{n+1} = u_n + \left[\frac{h}{2}\right][F(u_n) + F(v_{n+1})]$$

This is an example of a "predictor–corrector" method.

Adams–Bashforth methods. Let $q(t) = F(y(t))$ and let $F_j = F(y_j)$ for $0 \le j \le p$. Then

$$\int_{t_p}^{t_{p+1}} F(y(t))\,dt = \int_{t_p}^{t_{p+1}} q(t)\,dt$$

Set $Q(t)$ to be the unique polynomial of degree p such that $Q(t_j) = F_j$ for $0 \le j \le p$. Change variables by setting $t(s) = t_p + hs$, so that $dt = h\,ds$ and hence

$$\int_{t_p}^{t_{p+1}} q(t)\,dt = \int_0^1 Q(t(s))h\,ds + E$$

where E represents the error term. One can show that $E \leq Kh^{p+2}$ for some constant K which depends on bounds for the partial derivatives of F of order less than $p+2$. In fact,

$$E = \frac{\left(\frac{d}{dt}\right)^{p+1} F(y(\tau)) h^{p+2}}{(p+1)!} \int_0^1 (s)(s+1) \cdots (s+p)\, ds \qquad \text{for some } \tau \in [0, t_p]$$

Thus the local truncation error of the method is order $p+2$.

The Adams–Bashforth method is obtained by evaluating the integral of Q and ignoring the error term E. Let $g(s) = Q(t(s))$. Then g is the unique polynomial of degree p such that $g(-p+j) = F_j$ for $0 \leq j \leq p$. Thus, one can write

$$g(s) = \sum_{j=0}^{p} F_j L_j(s)$$

where $L_j(s)$ is the unique (Lagrange) polynomial of degree p such that

$$L_j(-p+i) = 0 \text{ for } i \neq j \text{ and } 0 \leq i \leq p \qquad \text{and } L_j(-p+j) = 1$$

Thus,

$$\int_0^1 Q(t(s))\, ds = \int_0^1 g(s)\, ds = \sum_{j=0}^{p} F_j \beta_j \qquad \text{where } \beta_j = \int_0^1 L_j(s)\, ds$$

The Adams–Bashforth methods are given by the formula:

$$u_{p+1} = u_p + h \sum_{j=0}^{p} F(u_j) \beta_j$$

Example: Suppose $p = 2$. Then $L_2(s) = (\frac{1}{2})[s^2 + 3s + 2]$, and $\beta_2 = 23/12$. The other β values can be similarly computed giving the three-step method:

$$u_3 = u_2 + \frac{h}{12}[5F(u_0) - 16F(u_1) + 23F(u_2)]$$

This method has local truncation error $O(h^4)$. The four-step method is

$$u_4 = u_3 + \frac{h}{24}[-9F(u_0) + 16F(u_1) - 59F(U_2) + 55F(u_3)]$$

The five-step method is

$$u_5 = u_4 + \frac{h}{720}[251F(u_0) - 1274F(u_1) + 2616F(u_2) - 2774F(u_3) + 1901F(u_4)]$$

Map Interpretation

Form a vector $U = (u_0, \ldots, u_p)$ (each u_j may itself be a vector) and define a map $A_h(u_0, \ldots, u_p) = (u_1, \ldots, u_{p+1})$ with

$$u_{p+1} = u_p + h \sum_{j=0}^{p} F(u_j)\beta_j$$

The general Adams–Bashforth method can be viewed as iterating this map.

The start-up problem: The startup values $(u_0, \ldots, u_p)$ must be calculated using a one-step method. The fourth-order Runge–Kutta method is often used. The step size might be reduced and more steps taken to get the starting values more accurately. Also, one might run the one-step method backward so that u_p is exactly the initial condition for the problem.

Derivation of Adams–Moulton Methods

These methods are derived in a fashion similar to the Adams–Bashforth methods. One approximates

$$\int_{t_p}^{t_{p+1}} y'(t)\,dt = \int_{t_p}^{t_{p+1}} F(y(t))\,dt \quad \text{with} \quad \int_{t_p}^{t_{p+1}} G(t)\,dt$$

where $G(t)$ is the unique polynomial of degree p such that $G(t_j) = F_j$ for $1 \leq j \leq p+1$. Change the time variable as before by setting $t(s) = t_p + hs$. Then

$$\int_{t_p}^{t_{p+1}} G(t)\,dt = \int_0^1 g(s)h\,ds$$

where g is the unique polynomial of degree p such that $g(-p+j) = F_j$ for $1 \leq j \leq p+1$.

One can write g in terms of Lagrange polynomials:

$$g(s) = \sum_{j=1}^{p+1} F_j P_j(s)$$

and one can express the integral of g by the formula

$$\int_0^1 g(s)h\,ds = h \sum_{j=1}^{p+1} F_j \gamma_j \quad \text{where } \gamma_j = \int_0^1 P_j(s)\,ds$$

This gives the general Adams–Moulton method:

APPENDIX B

$$u_{p+1} = u_p + h \sum_{j=1}^{p+1} F(u_j)\gamma_j$$

Examples: The two-step Adams–Moulton method is

$$u_3 = u_2 + \frac{h}{12}[-F(u_1) + 8F(u_2) + 5F(u_3)]$$

The three-step Adams–Moulton method is

$$u_4 = u_3 + \frac{h}{24}[F(u_1) - 5F(u_2) + 19F(u_3) + 9F(u_4)]$$

The four-step Adams–Moulton method is

$$u_5 = u_4 + \frac{h}{720}[-19F(u_1) + 106F(u_2) - 264F(u_3) + 6469F(u_4) + 251F(u_5)]$$

Since these methods are implicit, they are used together with Adams–Bashforth methods to generate predictor–corrector methods.

Example: Use a four-step Adams–Bashforth method as a "predictor" together with a four-step Adams–Moulton method as a "corrector":

$$v_4 = u_3 + \frac{h}{24}[-9F(u_0) + 16F(u_1) - 59F(u_2) + 55F(u_3)]$$
$$u_4 = u_3 + \frac{h}{24}[F(u_1) - 5F(u_2) + 19F(u_3) + 9F(v_4)]$$

Appendix C: Tangent Bundles, Manifolds, and Differential Forms

This is a quick review of some definitions from differential topology. Good references for further study are Guillemin and Pollack (1974) and Flanders (1963). Manifolds of dimension n are subsets of Euclidean spaces which are locally diffeomorphic to R^n. An open subset of a n-dimensional manifold which is diffeomorphic to an open subset of R^n is called a coordinate patch. The coordinates on R^n can be used as coordinates on the patch. The unit sphere in three-dimensional Euclidean space is an example of a two-dimensional manifold. At this time the global topology of manifolds will not be emphasized. Rather, we will work locally in a coordinate patch, and we will identify that patch with an open subset of R^n.

Definitions: Let U be an open subset of R^n. The set V^n will denote the vector space of all real column vectors of height n.

$T(U) = U \times V^n$ is called the *tangent bundle* of U.

$T_x(U) = x \times V^n$ is called the *tangent plane* to U at the point x in U.

The map $\pi: T(U) \to U; (x, v) \to (x)$ is called the *projection map* of the tangent bundle onto U.

The map $\exp: TU \to R^n$; $\exp(x, v) = x + v^t$ where v^t is the transpose of v is called the *exponential map* of the tangent bundle into R^n.

A continuous function $s: U \to T(U)$ is called a *section* of the tangent bundle when the composition of π with s is the identity map. A section of the tangent bundle of R^n is also called a *vector field* on R^n.

$T^*(U) = U \times R^n$ is called the *cotangent bundle* of U.

$\pi: T^*(U) \to U; (x, w) \to (x)$ projects the cotangent bundle onto U.

$T_x^*(U) = x \times R^n$ is called the *cotangent plane* to U at the point x in U.

Both tangent planes and cotangent planes are considered as vector spaces in the obvious way. The cotangent plane at x is viewed as the dual vector space to the tangent plane at x. A cotangent vector $w^* = (x, u)$ acts on a tangent vector $v = (x, u)$ to give the real number $\langle w^*, v \rangle = uv$ obtained by matrix multiplication.

Suppose that $A \subset R^n$ and $B \subset R^n$. A map $g: A \to B$ is said to be a C^r-*map* if there exist open sets U containing A and V containing B and a C^r function $G: U \to V$ such that, for each x in A, $G(x) = g(x)$. The map g is called a C^r-*diffeomorphism* if g is a homeomorphism and both g and its inverse are C^r maps.

Definition: If f is a differentiable map from an open set U into an open subset W of R^m, then associated with f there is a *tangent map*

$$Tf: T(U) \to T(W); (x, v) \to (f(x), Df(x)v)$$

PROPOSITION C.1: The tangent map of the composition of two maps is the composition of their tangent maps.

This is a consequence of the chain rule.

Definition: Let V be a real vector space. The *dual space* V^* of V is the vector space of all linear functions from V to R^1. If $T: V \to W$ is a linear transformation, then the *adjoint* linear transformation $T^*: W^* \to V^*$ is defined implicitly by the equation $T^*(w^*)(v) = w^*(T(v))$.

Notation: The dual space of V^n is R^n. The columns $\{E_j\}$ of the $n \times n$ identity matrix form the standard basis for V^n. The rows $\{e_j\}$ of the $n \times n$ identity matrix form the standard basis for R^n. A vector w^* in V^n acts on a vector v in R^n by matrix multiplication. Thus $w^*(v) = w^*v$ and therefore V^n is the dual space of R^n. A linear transformation A from V^n to V^m is identified with the matrix A whose jth column is $A(E_j)$. Thus $A(E_j) = AE_j$. The adjoint transformation associated with A has matrix equal to the transpose of A.

Definitions: Let W be a vector space. A function $\tau: W^k \to R^1$ is *multilinear* if it is linear in each variable. Such a function is called a *k-tensor*. A *k*-tensor τ is *alternating* if the output of τ changes sign whenever two variables are interchanged. Alternating *k*-tensors are called *exterior k-forms*.

If α is a *k*-tensor and β is an *r*-tensor, then their *tensor product* is

$$\alpha \otimes \beta(v^1, \ldots, v^k, w^1, \ldots, w^r) = \alpha(V^1, \ldots, v^k)\beta(w^1, \ldots, w^r)$$

If α is a *k*-tensor, then there is an exterior *k*-form associated with α:

$$\text{alt}(\alpha) = \frac{1}{k!} \sum (-1)^\sigma \alpha^\sigma \quad \text{where } \alpha^\sigma(v^1, \ldots, v^k) = \alpha(v^{\sigma(1)}, \ldots, v^{\sigma(k)})$$

The sum is taken over all permutations σ, and -1^σ is the sign of σ.

Notation: The vector space of all exterior k-forms on V is denoted by $\Lambda^k(V)$.

Example: The determinant of an $n \times n$ matrix can be viewed as an exterior n-form on V^n.

Definition: The *wedge product* of an a-form α with a b-form β is defined to be

$$\alpha \wedge \beta = \frac{(a+b)!}{a!b!} \text{ alt}(\alpha \otimes \beta)$$

Definition: If T is a linear transformation from V to W then the *adjoint* transformation on exterior k-forms is $T^*: \Lambda^k(W) \to \Lambda^k(V)$ with

$$T^*(\lambda)(v_1, \ldots, v_n) = \lambda(Tv_1, \ldots, Tv_n)$$

Definition: A *0-form* on U is a continuous real-valued function ϕ.

Definition: A *1-form* on U is a section λ of the cotangent bundle of U. Thus,

$$\lambda(x) = (x, (a_1(x), \ldots, a_n(x)))$$

Definition: The 1-form dx_j is defined by $dx_j(x) = (x, e_j)$. Thus, the 1-form λ above can be expressed as

$$\lambda = \sum_{j=1}^{n} a_j \, dx_j$$

Definition: If $\gamma: [a, b] \to U$ is a differentiable function, then the *line integral* of the 1-form λ over the curve γ is defined to be

$$\int_\gamma \lambda = \int_a^b \sum_{j=1}^{n} a_j(\gamma(t)) \langle dx_j(\gamma(t)), \dot{\gamma}(t) \rangle \, dt = \int_a^b \sum_{j=1}^{n} a_j(\gamma(t)) \dot{\gamma}_j(t) \, dt$$

Definition: The *wedge product* of two 1-forms dx_r and dx_s is defined by

$$dx_r \wedge dx_s(v, w) = dx_r(v) \, dx_s(w) - dx_r(w) \, dx_s(v)$$

Definition: A *k-form* on U is a section of $\Lambda^k(U)$.

THEOREM C.2: Any k-form μ can be uniquely expressed as

$$\mu = \sum_I a_I(x) \, dx_I$$

where $dx_I = dx_{i_1} \wedge \cdots \wedge dx_{i_k}$, and the sum is taken over all multi-indexes $I = (i_1, \ldots, i_k)$ with $1 \leq i_1 < \cdots < i_k \leq n$. The functions $a_I(x)$ are real-valued functions defined on U.

Definition: The *exterior derivative* of a differentiable function f (0-form) on U is

$$df(x) = \sum_{j=1}^{n} \frac{\partial f}{\partial x_j}(x)\, dx_j(x)$$

The exterior derivative of the k-form μ is the $k+1$ form

$$d\mu = \sum_I da_I \wedge dx_I$$

provided each of the functions a_I is differentiable.

Definition: Suppose that $h: R^m \to R^n$ is a differentiable function, and that α is a 1-form on R^n. The *pull-back* $h^*\alpha$ is the 1-form on R^m defined by the equation $h^*\alpha(v) = \alpha(Dh(v))$. Inductively, the pull back of a wedge product of forms is

$$h^*(\alpha \wedge \beta) = h^*\alpha \wedge h^*\beta$$

THEOREM C.3: The exterior derivative is characterized by the following four properties:

(1) $d(\alpha + \beta) = d\alpha + d\beta$.
(2) $d(\alpha \wedge \beta) = d\alpha \wedge \beta + (-1)^k \alpha \wedge d\beta$, where α is a k-form.
(3) $d(d(\alpha)) = 0$.
(4) $d(f^*\alpha) = f^* d\alpha$.

Each differential form can be integrated over an appropriate domain.

Definition: Suppose that $X \subset R^n$ and $Y \subset R^m$ are subsets of Euclidean spaces, and suppose that $\phi: X \to Y$ is a function. The function ϕ is said to be *of class C^r* if there exist open sets U and V containing X and Y, respectively, and a function $\Phi: U \to V$ whose partial derivatives of order r are continuous, and the function Φ agrees with ϕ on X. The function ϕ is said to be *smooth* if it is of class C^r for all r. The function ϕ is said to be a C^r *diffeomorphism* if it has an inverse, and the inverse is also of class C^r.

Definitions: A (closed) subset $X \subset R^n$ is a C^r *manifold of dimension d* if each point p in X has a neighborhood W contained in X which is C^r diffeomorphic to an open subset of R^d. If $\phi: W \to R^d$ is such a diffeomorphism, then the pair (W, ϕ) is called a *chart* (or *local coordinate system*) *at p*. One can consider

$\phi^{-1}: R^d \to R^n$ as a differentiable map, and define the *tangent plane to X at p* to be the subspace of the tangent plane to R^n at p defined by the formula $T_p(X) = T\phi^{-1}(T_a(R^d))$, where $a = \phi(p)$. One may also consider the tangent plane to X at p to consist of all tangent vectors in the tangent plane to R^n at p which arise as the tangent vector at p of a differentiable curve which is contained in X and passes through p. The tangent plane to X at p can be shown to be independent of the choice of the chart at p. The *tangent bundle* of X consists of all tangent planes to X. The dual space of the tangent plane to X at p is called the *cotangent plane to X at p* and is denoted $T_p^*(X)$. The *cotangent bundle* of X consists of all cotangent planes to X.

Examples: The unit sphere in R^{d+1} is an example of a d-dimensional manifold. Many manifolds arise in the following way: A collection of m real-valued differentiable functions on R^n is given together with m constants. The set X of all points of R^n, where each given function is equal to the corresponding constant, is a manifold provided that the gradients of the given functions are linearly independent at each point in X. It is a nice exercise involving the implicit function theorem to provide that X is a manifold of dimension $n - m$.

Appendix D: Symplectic Manifolds

Conservation laws such as conservation of energy or conservation of angular momentum for Hamiltonian systems of differential equations determine manifolds. For example, the set of states in phase space having constant energy is generally a manifold. Compact manifolds are interesting state spaces on which to study dynamics. To study conservative dynamics, it is necessary to define symplectic structures on even-dimensional manifolds.

Definition: A *symplectic structure* on a manifold M is a closed nondegenerate 2-form ω on M. Thus $d\omega = 0$, and for any $v \in T_p^*(M)$ there exists $w \in T_p^*(M)$ such that $\omega(v, w) = 1$. A manifold together with a symplectic structure is called a *symplectic manifold*.

Definition: If (M, ω) and (N, λ) are symplectic manifolds, a differentiable map $g: X \to Y$ is said to be a *symplectic map* if $g^*\lambda = \omega$.

The *standard symplectic manifold* for many applications is $(R^{2n}, dq \wedge dp)$. In this context, a map f of R^{2n} is symplectic if the matrix $Df(x)$ is a symplectic matrix for each x. Darboux's theorem (Abraham and Marsden, 1978) states that any symplectic manifold of dimension $2n$ is locally diffeomorphic to an open subset of R^{2n}, and the diffeomorphism is a symplectic map. Thus, local problems concerning symplectic manifolds and symplectic maps can be identified with problems concerning symplectic maps of $(R^{2n}, dq \wedge dp)$.

Definitions: An *exact symplectic manifold* is a pair (M, α) where α is a 1-form on M such that $(M, d\alpha)$ is a symplectic manifold.

The *standard exact symplectic manifold* is (R^{2n}, μ) where (q, p) are coordinates and

$$\mu = \sum_{k=1}^{n} p_k\, dq_k$$

Definition: An *exact symplectic map* of an exact symplectic manifold M is a map f such that $f^*\alpha - \alpha = dG$ for some differentiable real-valued function G defined on M.

If f is an exact symplectic map of V^{2n}, and γ is a curve in V^{2n}, then the definition requires that

$$\int_{f(\gamma)} \mu = \int_\gamma \mu + G(\gamma(1)) - G(\gamma(0))$$

PROPOSITION D.1: An exact symplectic map is a symplectic map.

Definition: A *submanifold* of a manifold M is a closed subset of M which is a manifold. A submanifold X of a symplectic manifold (M, ω) is a *symplectic submanifold* if the restriction of ω to X is a symplectic structure on X.

Definition: Let (M, ω) be a symplectic manifold of dimension $2d$. A submanifold L of M of dimension d is a *Lagrangian submanifold* if, as a 2-form on L, ω is identically zero.

PROPOSITION D.2: Let (M, ω) be a symplectic manifold. Then $M \times M$ is a symplectic manifold with the symplectic structure $\rho = \pi_2^*\omega - \pi_1^*\omega$ where π_1 and π_2 are the two projections of $M \times M$ onto M with $\pi_1(a, b) = a$ and $\pi_2(a, b) = b$.

Definition: The symplectic structure ρ in the above proposition is called the *product symplectic structure* on $M \times M$.

PROPOSITION D.3: If f is a smooth map of a manifold M, then the garph of f is a submanifold of $M \times M$. If (M, ω) is a symplectic manifold and if f is a smooth symplectic map, then the graph of f is a Lagrangian submanifold of $M \times M$, with the product symplectic structure $\pi_2^*\omega - \pi_1^*\omega$, where π_1 and π_2 are the projections of $M \times M$ onto M.

For a proof using local coordinates see Theorem 6.D.5.

Further Reading

A nice treatment of differential topology and differential forms can be found in Guillemin and Pollack (1974). An advanced modern treatment of Hamiltonian mechanics is presented in Abraham and Marsden (1978). Symplectic manifolds and symplectic geometry are discussed in Weinstein (1977), Weinstein (1981), McDuff (1990), and Hofer and Zehnder (1994).

Appendix E: Algebraic Topology

Topologists are interested in deciding whether or not two topological spaces are homeomorphic and in deciding whether or not two maps are homotopic. Algebraic topology provides tools for determining that spaces are not homeomorphic and that maps are not homotopic. Many scientists are not familiar with these tools, and a working knowledge requires at least a one-year graduate course in algebraic topology. Subsequently, I will state and use selected results. A source book that I have found useful is Spanier (1966).

Definition: Two maps f and g from a metric space X to a metric space Y are *homotopic* if there exists a continuous function $H : X \times [0, 1] \to Y$ such that

$$H(x, 0) = f(x) \text{ for all } x \in X \quad \text{and} \quad H(x, 1) = g(x) \text{ for all } x \in X$$

Definition: Let X be a path-connected metric space and let x_0 be a point in X. A *loop* in X is a map $\alpha \colon [0, 1] \to X$ with $\alpha(0) = x_0 = \alpha(1)$. Two loops α and β can be "multiplied" together to form a loop γ, where $\gamma(s) = \alpha(2s)$ for $0 \le s \le 0.5$ and $\gamma(s) = \beta(2s - 1)$ for $0.5 \le s \le 1$. There is an equivalence relation on the set of all loops defined by $\alpha \simeq \beta$ provided there is a homotopy $H : [0, 1] \times [0, 1] \to X$ such that $H(s, 0) = \alpha(s)$ and $H(s, 1) = \beta(s)$ for all s, and $H(0, t) = x_0 = H(1, t)$. Thus, there is a one-parameter family of loops joining α and β. The *fundamental group* $\pi_1(X)$ of the space X is defined to be the set of equivalence classes of all loops. Multiplication is defined by $[\alpha]^*[\beta] = [\alpha^*\beta]$. It is a theorem that multiplication is well defined, and that the fundamental group is independent of the choice of the base point x_0 up to isomorphism.

Example: Consider the set $C^0(S^1)$ of all maps from the unit circle to itself. This is a metric space with metric defined by $d(f, g) = \sup\{d(f(x), g(x))\}$. The supremum is taken over all points x of the circle. Maps which are less than distance 2 apart (2 is the diameter of the unit circle) are homotopic. The relation on maps of being homotopic is an equivalence relation. The equivalence classes can be counted by what is called the winding number of a map. The *winding number* of a map is the number of times the image of the circle is wrapped counterclockwise around the unit circle. Clockwise rotation is counted as negative. It is a theorem that the homotopy equivalence classes are in one-to-one correspondence with the integers via the winding number. Consider next a map G of the plane, which has an isolated fixed point p. Choose a circle C centered at p of radius r such that p is the only fixed point of G inside or on this circle. For each x belonging to C define a unit vector $w(x) = (G(x) - x)/|G(x) - x|$. Finally, define a map ω of the unit circle by mapping it first to C and then back to the unit circle by w. The map ω is given by the formula $\omega(u) = (G(x) - x)/|G(x) - x|$ with $x = p + u/r)$. The *index* of the fixed point p is defined to be the winding number of the map ω. It is a theorem that the index is independent of the construction, and that if the fixed point continues as an isolated invariant set of a nearby map of the plane, then its index with respect to this map is unchanged.

Homology theories assign groups to spaces and homomorphisms to maps. This assignment is technically called a "function." There is also a group of coefficients involved in the construction, which we will take to be the real numbers. Thus, if X and Y are metric spaces, and f is a map from X to Y, for each nonnegative integer k there is a real vector space (an Abelian group) $H_k(X)$ called the kth *homology group of X* assigned to the space X, and there is a linear transformation (homomorphism) $f_*: H_k(X) \to H_k(Y)$ associated to the map f. If g is a map which is homotopic to f, then $f_* = g_*$. Further, the linear transformation assigned to the composition of maps is the composition of the linear transformation assigned to each. For example, if e is a map from Y to a metric space Z, then $(ef)_* = e_* f_*$.

Cohomology theories associate to a space the dual spaces of its homology groups, and associate to a map between spaces the linear transformation of their dual spaces naturally defined in terms of the adjoint to the linear transformation of their homology groups. Thus, $H^k(X)$ denotes the dual space of the vector space $H_k(X)$, and $f^*: H^k(Y) \to H^k(X)$ is the induced transformation. Note that f^* reverses direction relative to the map f.

From now on we use the Alexander–Spanier cohomology theory (see Spanier, 1966, pp. 306–323). This theory has the following important "continuity," "excision," and "duality" theorems.

THEOREM E.1 (Continuity): Let $\{X_j\}$ be a nested sequence of compact subsets of a metric space X directed downward by inclusion, and let Z denote the intersection of these sets. Then the inclusion maps induce an isomorphism $H^k(Z) \approx \text{Directlimit } H^k(X_j)$.

Note: Homology and cohomology theories also assign vector spaces to pairs of spaces. The pair of spaces (X, A) indicates that A is a subset of X.

THEOREM E.2 (Excision): Let X be a compact metric space, let A be a compact subset of X, and let U be an open subset of X which is contained in A. Then the inclusion map $i: (X - U, A - U) \to (X, A)$ induces an isomorphism of cohomology groups $i^*: H^k(X, A) \to H^k(X - U, A - U)$.

THEOREM E.3 (Duality): Let M be an oriented manifold of dimension n, and let A be a compact subset of M. Then $H^k(A) \approx H_{n-k}(M - A)$. (To be precise, the vector space $H_{n-k}(M - A)$ is the singular homology group in dimension $n - k$ with real coefficients for the space $M - A$.)

There is one further theorem that relates cohomology groups to topological dimension.

THEOREM E.4: If X is a compact metric space and $H^k(X)$ has positive dimension as a vector space, then the topological dimension of X is at least k.

References

Abraham, R. and J. E. Marsden (1978). *Foundations of Mechanics*. Benjamin, Reading, Mass.

Arnol'd, V. I. (1978). *Mathematical Methods of Classical Mechanics*. Springer-Verlag, New York.

Arnol'd, V. I. (ed.) (1988). *Dynamical Systems III*. Springer-Verlag, New York.

Arrowsmith, D. and C. Place (1990). *An Introduction to Dynamical Systems*. Cambridge University Press, New York.

Barge, M. and R. Swanson (1990). "Pseudo-Orbits and Topological Entropy," *Proc. AMS*, **109**, 559–566.

Benedicks, M. and L. Carleson (1991). "The Dynamics of the Hénon map," *Ann. Math.*, **133**, 73–169.

Benedicks, M. and L. Young (1993). "SBR Measures for Certain Hénon Maps," *Invent. Math.*, **112**, 541–576.

Billingsley, P. (1965). *Ergodic Theory and Information*. Wiley, New York.

Bowen, R. (1975). *Equilibrium States and the Ergodic Theory of Anosov Diffeomorphisms*. Springer-Verlag, New York.

Bowen, R. and D. Ruelle (1975). "The Ergodic Theory of Axiom A Flows," *Invent. Math.*, **29**, 181–202.

Conley, C. (1975). *Hyperbolic Sets and Shift Automorphisms*. Springer-Verlag, New York.

Conley, C. (1978). *Isolated Invariants Sets and the Morse Index*. American Mathematical Society, Providence, R.I.

Conley, C. and R. Easton (1971). "Isolated Invariant Sets and Isolating Blocks," *Trans. AMS*, **158**, 35–59.

Devaney, R. (1986). *An Introduction to Chaotic Dynamical Systems*. Benjamin/Cummings, Menlo Park, Calif.

Devaney, R. and L. Keen (1989). *Chaos and Fractals*. American Mathematical Society, Providence, R.I.

Easton, R., J. Meiss and S. Carver (1993). "Exit Times and Transport for Symplectic Twist Maps," *Chaos*, **3**, 153–165.

Easton, R. W. (1985). "Trellises Formed by Stable and Unstable Manifolds in the Plane," *Trans. AMS*, **294**, 719–732.

REFERENCES

Easton, R. W. (1989). "Isolating Blocks and Epsilon Chains for Maps," *Physica D*, **39**, 95–110.

Easton, R. W. (1991). "Transport Through Chaos," *Nonlinearity*, **4**, 583–590.

Easton, R. W. (1993). "Transport of Phase Space Volume Near Isolated Invariant Sets," *J. Dyn. Diff. Equations*, **5**, 529–536.

Feingold, M., L. Kadanoff, and O. Piro (1989). *Transport of Passive Scalars: KAM Surfaces and Diffusion in Three-dimensional Liouvillian Maps*. Kluwer Academic Publishers, Dordrecht.

Flanders, H. (1963). *Differential Forms with Applications to the Physical Sciences*. Academic Press, New York.

Ge, Z. (1991). "Equivalent Symplectic Differences Schemes and Generating Functions," **49**, 376–386.

Guckenheimer, J. and P. Holmes (1983). *Nonlinear Oscillations, Dynamical Systems, and Bifurcation of Vector Fields*. Springer-Verlag, New York.

Guckenheimer, J., J. Moser, and S. Newhouse (1978). *Dynamical Systems*. Birkhauser, Boston, Mass.

Guillemin, V. and A. Pollack (1974). *Differential Topology*. Prentice-Hall, London, UK.

Gumowski, I. and C. Mira (1980). *Recurrences and Discrete Dynamic Systems*. Springer-Verlag, New York.

Hénon, M. (1969). "Numerical Study of Quadratic Area-Preserving Mappings," *Q. J. Appl. Math.*, **27**, 291–312.

Herman, M., R. McGeHee, J. Moser, and E. Zender (1988). *Charles Conley Memorial Volume*. Cambridge University Press, New York.

Hofer, H. and E. Zehnder (1994). *Symplectic Invariants and Hamiltonian Dynamics*. Birkhauser, Boston, Mass.

Kurland, H. L. (1996). *Intersection Pairings on Conley Indices*. American Mathematical Society, Providence, R.I.

Lomeli, H. and J. D. Meiss (1997). "Quadratic Volume Preserving Maps," preprint.

MacKay, R. S. and J. D. Meiss (ed.) (1987). *Hamiltonian Dynamical Systems: A Reprint Selection*. Adam Hilger, London, UK.

MacKay, R. S., J. D. Meiss and L. I. C. Percival (1984). "Stochasticity and Transport in Hamiltonian Systems," *Phys. Rev. Lett.*, **52**, 697–700.

Mane, R. (1987). *Ergodic Theory and Differentiable Dynamics*. Springer-Verlag, New York.

McDuff, D. (1990). "Elliptic Methods in Symplectic Geometry," *Bull. AMS*, **23**, 311–358.

McGehee, R. (1973). "A Stable Manifold Theorem for Degenerate Fixed Points with Applications to Celestial Mechanics," *J. Diff. Equations*, **14**, 70–88.

McGehee, R. (1992). "Attractors for closed relations on compact Hausdorff spaces." *Indiana U. Math. J.*, 1165–1209.

Meiss, J. D. (1992). "Symplectic Maps, Variational Principles, and Transport," *Rev. Mod. Phys.*, **64**, 795–848.

Meyer, K. and R. Hall (1991). *Introduction to Hamiltonian Dynamical Systems and the N-Body Problem*. Springer-Verlag, New York.

Mischaikow, K. (1995). *Conley Index Theory*. Lecture Notes in Mathematics No. 1609, Springer-Verlag, New York.

Moser, J. (1977) "Proof of a Generalized Form of a Fixed Point Theorem due to G. D. Birkhoff," *Lecture Notes in Mathematics*, **597**, 464–494.

REFERENCES

Moser, J. (1973). *Stable and Random Motions in Dynamical Systems.* Princeton University Press, Princeton, N.J.
Moser, J. (1994). "On Quadratic Symplectic Mappings," *Math. Zeitschrift*, **218**, 417–430.
Moser, J. K. (1968). "Lectures on Hamiltonian Systems," *Mem. Am. Math. Soc.*, **81**, 1–60.
Mrozek, M. (1990). "Leray Functor and Cohomology Conley Index for Discrete Dynamical Systems," *Trans. AMS*, **318**, 149–178.
Mrozek, M. and K. P. Rybakowski (1991). "A Cohomological Conley Index for Maps on Metric Spaces," *J. Diff. Equations*, **90**, 143–171.
Poincaré, H. (1899). *Les méthodes nouvelles de la mécanique céleste.* Gauthier-Villar, Paris.
Robin, J. W. and D. Salamon (1988). "Dynamical Systems, Shape Theory, and the Conley Index," *Ergodic Theor. Dyn. Syst.*, **8**, 373–393.
Robinson, C. (1995). *Dynamical Systems.* CRC Press, Ann Arbor, Mich.
Rom-Kedar, V. (1994). "Homoclinic Tangles-Classification and Applications," *Nonlinearity*, **7**, 441–473.
Rom-Kedar, V. and S. Wiggins (1988). "Transport in Two-Dimensional Maps," *Arch. Rational Mech. Anal.*, **109**, 239–298.
Ruelle, D. (1989). *Elements of Differentiable Dynamics and Bifurcatuion Theory.* Academic Press, San Diego, Calif.
Rybakowski, K. (1987). *The Homotopy Index and Partial Differential Equations.* Springer-Verlag, New York.
Shub, M. (1987). *Global Stability of Dynamical Systems.* Springer-Verlag, New York.
Siegel, C. L. and J. Moser (1971). *Lectures on Celestial Mechanics.* Springer-Verlag, New York.
Smoller, J. (1983). *Shock Waves and Reaction–Diffusion Equations.* Springer-Verlag, New York.
Spanier, E. (1966). *Algebraic Topology.* McGraw-Hill, New York.
Walters, P. (1981). *An Introduction to Ergodic Theory.* Springer-Verlag, New York.
Weinstein, A. (1977). *Lectures on Symplectic Manifolds.* American Mathematical Society, Providence, R.I.
Weinstein, A. (1981). "Symplectic Geometry," *Bull. Am. Math. Soc.*, **5**, 1–13.
Wiggins, S. (1992). *Chaotic Transport in Dynamical Systems.* Springer-Verlag, New York.
Yoshida, H. (1993). "Recent Progress in the Theory and Application of Symplectic Integrators," *Celestial Mechanics and Dynamical Astronomy*, **56**, 27–43.

Index

action function, 111
Adams–Bashforth methods, 138
Adams–Moulton methods, 140
adjoint linear transformation, 143, 144
affine man, 38
Alexander–Spanier cohomology, 150
almost everywhere, 128
Arnold's circle maps, 13
array of blocks, 88
asymptotically stable, 74
attracting fixed point, 6, 11
attracting set, 73
attractor block, 74
Aubry–Mather sets, 13

ball of radius r, 132
basic sets, 27, 83
basin of attraction, 74
bicompact, 133
block vertex, 88
Borel measure, 127
boundary, 132
branched horseshoe map, 78

Cantor set, 134
Cauchy sequence, 131
chain entropy, 70
chain equivalence class, 27
chain recurrent set, 27
chain stable, 74
characteristic function, 128
chart, 145
class C^r, 35
closed, 132

closure, 132
coisotropic, 106
compact, 133
complete, 132
confining set, 20
Conley decomposition theorem, 28
Conley index, 94, 98, 99
connected, 133
connection, 90
continuation, 96
continuous, 132
contraction map, 135
converges, 132
cotangent bundle, 142
cotangent plane, 142
C^r-diffeomorphism, 143

differentiable stable manifold theorem, 41
differential, 35
direct limit, 99
direct sum, 99
directed graph, 30, 88
disconnected, 133
discrete dynamical system, 19
dual space, 143
Duffing's equation, 15
dynamical system, 18

elliptic fixed points, 14
epsilon chains, 25
ergodic, 129
Euler–Lagrange equations, 110
eventually periodic, 20
exact symplectic manifold, 147

exact symplectic map, 112
excision, 151
exit set, 32
exit threshold set, 32
exit time, 32
exit time decompositions, 32, 40
expansive, 50
exponential map, 142
exterior, 132
exterior derivative, 145
exterior k-forms, 143

filtration, 85
fixed point, 5
flow, 19
forward invariant set, 20
fractals, 135
full orbit, 20
fundamental group, 149
fundamental segment, 61

generating function, 114, 115
geodesic, 110
graphical analysis, 5

Hamiltonian system, 107
Hartman–Grobman theorem, 52
Hausdorff metric, 134
Hénon maps, 9, 63
heteroclinic point, 90
Heun method, 138
holonomic force, 107
homeomorphism, 132
homoclinic points, 11
homoclinic tangle, 60
homology group, 150
homotopic, 149
horizontal pair, 42, 48
horizontal slice, 42, 48
horseshoe map, 54
hyperbolic fixed point, 36
hyperbolic invariant set, 58

immersion, 60
index map, 93
index of the fixed point, 150

index space, 93
initial segment, 61
interior, 132
invariant curves, 12
invariant measure, 127
invariant set, 20
isolated invariant set, 73
isolating block, 73
isolating region, 73
isotropic, 106
itinerary, 56
itinerary map, 56, 84

Jacobian matrix, 10, 15

k-tensor, 143

Lagrangian, 110
Lagrangian submanifold, 113, 148
leapfrog method, 118
Legendre transformation, 111
limit point, 132
limit stable, 74
line integral, 144
linearization procedure, 37
Liouville's theorem, 109
Lipschitz, 25
Lipschitz stable manifold theorem, 41
local continuation, 96
local coordinate system, 145
local stable set, 77
locally compact, 133
locally conjugate, 51
logistic map family, 4

Mandelbrot set, 33
manifold, 145
map, 19
matrix norm, 38
maximal invariant set, 23
measure, 126
measure space, 126
Melnikov, 119
metric space, 131
midpoint rule, 137
Morse set, 85

INDEX

multilinear, 143

neighborhood, 132
network of blocks, 87
nonresonance conditions, 121
normal forms, 121

omega limit point, 23
omega limit set, 21
open, 132
orbit, 19

partial derivatives, 35
periodic, 20
persistent, 101
phase portrait, 49
Poincaré map, 15
Poincaré recurrence theorem, 127
potential energy, 107
preorbit, 20
primary homoclinic point, 61
product symplectic structure, 148
pull back, 145

quadratic maps, 14
quasi-attractor, 76
quotient space, 93
quotient topology, 93

R-stable and R-unstable manifolds, 62
repelling fixed point, 6, 11
resonance zone, 62
retraction, 71
rotation number, 13

saddle point, 11, 36
semi-conjugate, 72
separatrix movement, 118
shadowing property, 50
shift automorphism, 56, 84
short filtration, 85
sigma-algebra, 126
skew orthogonal, 104
smooth, 145
solenoid, 74
stable manifold, 11

stable set, 23, 32
stable and unstable manifolds, 60
stable and unstable orderings, 61
stack of blocks, 86
standard map family, 12
states, 18
strings, 65
structurally stable, 49
submanifold, 148
subsequence, 131
subshift of finite type, 84
symbol space, 56, 83
symbolic dynamics, 56
symplectic basis, 103
symplectic group, 104
symplectic integrators, 117
symplectic manifold, 147
symplectic map, 14, 104, 108, 147
symplectic matrix, 104
symplectic structure, 103, 147
symplectic submanifold, 147
symplectic subspace, 105
symplectic vector space, 103

tangent bundle, 142, 146
tangent map, 49, 143
tangent plane, 142
tensor product, 143
topological entropy, 70
topological stable manifold theorem, 39
topologically conjugate, 36
transverse homoclinic point, 61
trapezoid method, 138
trapping region, 74
trellis, 59, 63

uniformly continuous, 133
unstable manifold, 11

variational principles, 110
vector field, 142
vertical pair, 42

wandering point, 34
wedge product, 143
winding number, 150